ROUTLEDGE LIBRARY EDITIONS:
POLITICAL GEC

T0230889

Volume 6

NATIONALISM SELF-DETERMINATION AND POLITICAL GEOGRAPHY

NATIONALISM SELF-DETERMINATION AND POLITICAL GEOGRAPHY

Edited by
R.J. JOHNSTON, DAVID B. KNIGHT AND
ELEONORE KOFMAN

Routledge
Taylor & Francis Group

LONDON AND NEW YORK

First published in 1988

This edition first published in 2015
by Routledge
2 Park Square, Milton Park, Abingdon, Oxon, OX14 4RN

and by Routledge
711 Third Avenue, New York, NY 10017

Routledge is an imprint of the Taylor & Francis Group, an informa business

British Library Cataloguing in Publication Data
A catalogue record for this book is available from the British Library

ISBN: 978-1-138-80830-0 (Set)
eISBN: 978-1-315-74725-5 (Set)
ISBN: 978-1-138-80985-7 (Volume 6)
eISBN: 978-1-315-74976-1 (Volume 6)
Pb ISBN: 978-1-138-80987-1 (Volume 6)

Publisher's Note
The publisher has gone to great lengths to ensure the quality of this reprint but
points out that some imperfections in the original copies may be apparent.

Disclaimer
The publisher has made every effort to trace copyright holders and would
welcome correspondence from those they have been unable to trace.

Printed and bound by CPI Group (UK) Ltd, Croydon, CR0 4YY

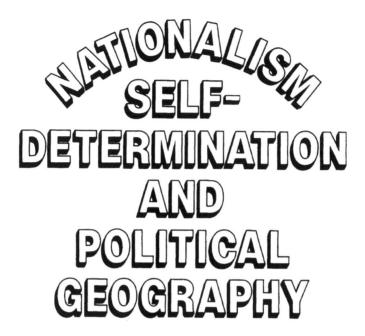

NATIONALISM SELF-DETERMINATION AND POLITICAL GEOGRAPHY

Edited by R.J. JOHNSTON, DAVID B. KNIGHT
and ELEONORE KOFMAN

CROOM HELM
London • New York • Sydney

© 1988 R.J. Johnston, David B. Knight and Eleonore Kofman
Croom Helm Ltd, Provident House, Burrell Row,
Beckenham, Kent, BR3 1AT
Croom Helm Australia, 44-50 Waterloo Road,
North Ryde, 2113, New South Wales

Published in the USA by
Croom Helm
in association with Methuen, Inc.
29 West 35th Street
New York, NY 10001

British Library Cataloguing in Publication Data

Nationalism, self-determination and
    political geography.
    1. Nationalism
    I. Johnston, R.J. II. Knight, David
    III. Kofman, Eleonore
    320.5′4   JC311
    ISBN 0-7099-1480-6

Library of Congress Cataloguing-in-Publication Data
ISBN 0-7099-1480-6

Printed and bound in Great Britain by Mackays of Chatham Ltd, Kent

CONTENTS

Preface

1. Nationalism, Self-Determination and the World
   Political Map: An Introduction
   R.J. Johnston, David B. Knight and Eleonore
   Kofman   1

2. Nationalist Ideology and Territory
   James Anderson   18

3. Nation-State Building in a "Newly-Industrialized
   Country": Reflections on the Brazilian
   Amazonia Case
   Bertha K. Becker   40

4. Marxism and Self-Determination: the Case of
   Burgenland, 1919
   Andrew Burghardt   57

5. Toward a Geography of Peace in Africa: Redefining
   Sub-State Self-Determination Rights
   Josiah A.M. Cobbah   70

6. National Integration Problems in the Arab World:
   the Case of Syria
   Alasdair Drysdale.   87

7. Problems in Combining Labour and Nationalist
   politics: Irish Nationalists in Northern
   Ireland
   Clive Hedges   102

8. Self-Determination for Indigenous Peoples:
   the Context for Change
   David B. Knight.   117

9. Evolving Regionalism in Linguistically Divided
   Belgium
   Alexander B. Murphy   135

10. Nationalism, Social Theory and the Israeli/
    Palestinian Case
    Juval Portugali.   151

11. Ethnoregional Societies, 'Developed Socialism'
    and the Soviet Ethnic Intelligentsia
    Graham Smith   166

12. The Occurrence of Successful and Unsuccessful
    Nationalisms
    H. van der Wusten.   189

13. Minority Nationalist Historiography
    Colin H. Williams.   203

About the Authors   222

Papers Prepared for the San Sebastian Conference. .   223

Index   226

PREFACE

Political geography is enjoying a welcome resurgence, as geographers of all subdisciplinary persuasions realise the need for a full appreciation of the nature and role of the state in human affairs. To provide a focus for that work, steps were taken in the early 1980s, led by John House, to promote the establishment of a Commission on Political Geography within the International Geographical Union. A major international conference was convened in Oxford by John House in 1983, just six months prior to his untimely death, at which a proposal for a Commission was written. The focus of its work was to be an understanding of the World Political Map, and I had the honour of election as Chairman-designate.

At about the same time, the Executive Committee of the International Geographical Union initiated a new organisational category of Study Groups, which were to have a limited life of four years during which the case for establishment of a full Commission could be made. Consequently the case was made for establishment of a Study Group on the World Political Map, and this was agreed at the Paris International Geographical Congress in 1984.

During the four-year life of the Study Group a full programme of meetings and conferences has been arranged in several parts of the world, each focusing on a particular topic. It is hoped that a number of publications will stem from these meetings, helping to inform the geographical community at large of our work and the contributions that it is making.

The first book to be published from the Group's meetings was a selection of papers from the 1983 Oxford conference (P.J. Taylor and J.W. House, editors, <u>Political Geography: Recent Advances and Future Directions</u>. Croom Helm, London, 1984). This volume is the third to be produced by the group (the second was Gerald Blake, editor, <u>Maritime Boundaries and Ocean Resources</u>. Croom Helm, London, 1987). It is derived from a conference that the group held in San Sebastian, Spain, in August 1986, as part of the Regional Conference of the International Geographical Union which focused on Barcelona. Only a selection of the papers presented there on these important topics is included in the book: the full list of papers presented is given at the end of the volume.

On behalf of the Study Group, I am grateful to all those who have made the production of this book possible. In particular, thanks are due to F. Gomez Pineiro, Juan Antonio Saez, and Lide Sanches, who did so much to make the meeting in San Sebastian a success, to the contributors to the formal and informal discussions then, and to the authors for collaborating in the production of this volume.

R.J. Johnston   **May 1987, Sheffield**

Chapter 1

NATIONALISM, SELF-DETERMINATION AND THE WORLD POLITICAL MAP:
AN INTRODUCTION

R.J. Johnston, David B. Knight and Eleonore Kofman

In the 1970s it was commonplace to open a discussion of
nationalism with remarks about the paucity of literature,
and the fact that many interpretations of nationalism saw it
as an irrational force rooted in atavistic sentiments.
However, by the mid-1980s such a statement, though still
occasionally uttered, would not be at all accurate. In the
intervening years there have been numerous studies of
nationalism and self-determination. There have been
theoretical (B. Anderson, 1983; Gellner, 1983; Munck, 1986;
A.D. Smith 1971, 1981), comparative (Breuilly, 1982;
Sathyamurthy, 1983; Tiryakian and Rogowski, 1985) and
individual case studies. The last are too numerous to
mention; many of them, like the comparative studies, have
contributed, through an examination of particular national-
ism(s), to more general debates (on internal colonialism,
Hechter, 1975; uneven development, Nairn, 1977; primordialism and territorial aspects, Linz, 1985; and self-determination, Knight and Davies, 1987).
      Geographers have not been absent from this growing
body of literature (e.g. J.Anderson, 1986; Agnew, 1984,
1987; Blaut, 1986; Knight, 1982; McLaughlin, 1986; G. Smith,
1979; Williams, 1982, 1985; Orridge and Williams, 1982). For
Agnew (1987), the way that nationalism has been defined and
examined in history and social science has not allowed for
geographical analysis. Yet Williams (1985, pp. 340-50)
lists five areas in which geographers have extended, or
could extend, the analysis of nationalism: 1) the national
construction of social space; 2) uneven development and
nationalism; 3) the secular intelligentsia; 4) structural
preconditions and triggering factors; and 5) ecological
analysis.
      Before we embark on an exploration of what geographers
can offer to the study of nationalism, especially the
appreciation of territory as the basis and political
resource of nation-building in a world of states, we shall

1

first clarify what is meant by nationalism and self-determination.

Varieties of Nationalism

Given its varieties and complexity it is not surprising that definitions of nationalism have attempted to combine its multiple aspects. Each of these aspects is constructed and does not constitute a 'natural' element. Thus Emerson (1960) defined nationalism as

> a community of people who feel that they belong together in the double sense that they share deeply significant elements of a common heritage and that they have a common destiny for the future ... [it] has become the body which legitimizes the state (pp. 95-6).

The implication is that a nation is historically derived - the past produces the present and leads to the future. More recent definitions have tended to emphasise the relationship between nationalism and the acquisition and sustaining of a state of one's own (Worsley, 1984, p. 248). It is increasingly recognised that nationalism is a political principle, that its ability to achieve its objectives involves practical politics (see Hedges chapter 7), and that the constituent cultural elements are not given and fixed for all time. For example, Basque nationalism has shifted from the primordial basis of its earliest form at the end of the nineteenth century to a more territorial form in its recent manifestations (Linz, 1985).

The political translation of aspirations can also be clearly seen in demands for self-determination, defined as the 'right of a group with a distinctive politico-territorial identity to determine its own destiny' (Knight, 1984). As both Cobbah (chapter 5) and Knight (chapter 8) stress, it is very difficult for minority groups to bypass the jurisdiction of an existing state, and thus to "dismember or impair, totally or in part", the existing state (United Nations Charter, quoted by Knight, chapter 8). What has been done in the process of colonisation, cannot easily be undone through decolonisation. The political implications of nationalism in a global system of states have made it virtually impossible for nationalism to be conceived as cultural autonomy in the way that, for example, Otto Bauer had advocated at the beginning of the twentieth century for minorities in the Habsburg Empire (Burghardt, chapter 4; Munck, 1986 pp. 39-41).

Typologies of nationalism have generally focused on the presence or absence of a state and whether the nation has still to be constructed. A.D. Smith (1971) has produced a division into pre- and post-independence, while Breuilly (1982) situates nationalist movements historically, in a world either without or of states. What is most significant nowadays is that the state is the accepted model of territorial organisation and the key determinant in the world political map. Portugali (chapter 10) argues that nationalism, and consequently the nation-state, has been elevated to the generative social order, that is, when a given configuration of events predominates and other events or configurations are subjected to this newly established order. Nationalism is the attempt to make the cultural congruent with the political: to each nation, in whatever terms it is defined, there should correspond a state: and in existing states, a nation should be forged, if it does not already exist.

It is in relation to these strategies that the second criterion of many typologies has focused. Thus Breuilly (1982) distinguishes between unification, reform and separatist nationalisms; for Orridge and Williams (1982), it is state nationalism, unification, irredentist, autonomist and separatist nationalisms. J. Anderson (1983, 1986) combines the various formative roles and types of strategies to arrange a sequence of nationalism and state formation (see figure 1.1).

Nationalisms, History and Territory

Throughout the historical sequence from the original types of nationalism via those of European settlement and anti-and post-colonial nationalism to, most recently, the old state separatisms, the processes of aligning the nation, based on cultural elements, with the sovereign state have brought into question the delimitation of boundaries and the appropriation of territory and history. Nationalist ideologies have sought to interpret the occupation and control of space, both in the past and as a plan for the future. This Janus-like quality, looking both ways to the past and the future, is equally relevant for states which seek to create and reinforce a sense of nationhood, and for nationalisms that oppose existing states in their attempt to carve out autonomous identities. Opposition to an existing state may occur both in the aftermath of decolonisation in overseas European-settler states and in older European states.

Furthermore, in making claims to nationhood and territory, the writing of geography and history are at stake. These are constantly in flux, for a nationalism, at

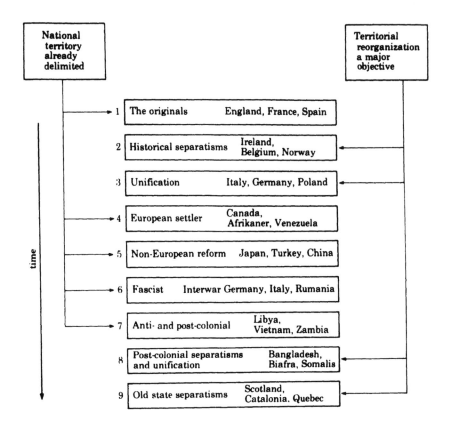

Figure 1.1. Varieties of nationalism (Source: J. Anderson, 1986, p. 122: reproduced with permission)

one time in opposition, may if successful acquire a state, whose legitimacy it, in turn, needs to consolidate, sometimes against a newly emergent separatism. Controlling the national past is a means by which other national, often deemed subordinate, pasts can be eliminated. It is a means by which consciousness is structured and experiences are colonised (Wright, 1985, p. 142). Unity has to be achieved through the 'historicity of a territory and territorialisation of a history' (Poulantzas, 1978, p. 114). Thus minority historiography (see Williams, ch. 13) is vital in questioning the contours of the state and offering alternative definitions of what is 'normal, appropriate or possible' (Wright, 1985, p. 42).

The significance of territory and territoriality in this process of nation-state building, though acknowledged by many writers, has been much more fully elaborated by geographers. Territory is bounded space, that is, 'a very substantial, material, measurable, and concrete entity' although it 'is [also] the product and indeed the expression of the psychological features of human groups' (Gottmann, 1973, p. 15). Nationalists desirous of matching their distinctive group identity with the claimed or held territory would accept the contention that

> territory ... is space to which identity is attached by a distinctive group who hold or covet that territory and who desire to have full control over it for the group's benefit (Knight, 1982, p. 526).

This attachment and desire can be an expression of territoriality. Sack (1983, p. 56) presents territoriality as

> the attempt by an individual or groups (x) to influence, affect or control objects, people and relationships (y) by delimiting and asserting control over a geographic area.

He suggests ten reasons (or tendencies) why territoriality is a viable means of exercising control. One of the most important for the present context is that it reifies power, identifying it with a place rather than social relations. In other words, territory can be used ideologically, to promote certain interests which require social control by associating them with a place within which that control is exercised (and recognised under the doctrine of sovereignty), thereby legitimating the control by obscuring its real nature.

In the original state nationalisms, the boundaries were largely laid down well before an identification with a sense of nationhood. In theory, such nationalisms involved the development of 'the economic, intellectual, and military resources of population ... [binding] them together in an ever tighter network of communications and complementarity based on an ever broader and more thorough participation of the masses of the populace' (Deutsch, 1953, p. 184). In reality, there was no inevitability about the nature or rapidity of such developments. Indeed national identities did not emerge smoothly nor without considerable effort expended by state institutions and the gradual creation of unified (state-wide) market economies. McLaughlin (1986, p. 301) concluded that

> National hegemony met widespread resistance ... and national unification and the drive to create central-ized national government were constantly in conflict with demands for local regional autonomy and self-government well into the present century.

In France, for example, it has been argued that the inclu-sion of the peasantry into the national domain did not occur until the end of the nineteenth century (Weber, 1976); the bourgeois Republic and its institutions were resisted by reactionary provincial forces until then. It was therefore considered to be of the utmost importance to imbue future generations with a solid sense of France's territorial possessions and history, as depicted in a popular school text of the period, entitled <u>Le Tour de France de Deux Enfants</u> (Sivignon, 1981, pp. 273-4). And in Spain, for instance, the two most developed regional societies con-tested, in their different ways, the correspondence of the state with a single nation. National unity was one of the key principles fought over in the bitter Civil War. Today, this issue is still not resolved in Euzkadi (Basque coun-try), where the Spanish state and moderate and radical nationalists disagree over the boundaries and the degree of sovereign power which should be exercised by the Basque political body (J. Anderson, 1985; Juaregui, 1987).

Discontinuous and lengthy national integration does not typify these original nation-states only. In certain states created by European settlement, as with Brazil, vast frontier regions have only recently been fully incorporated into such societies with their integration into the world economy. It is through the activities of the state in such frontier regions, with weakly developed civil societies, that the struggle for control is taking place. The frontier is both a physical and a symbolic space of production and reproduction of national identity (Becker, chapter 3).

A large number of states have, of course, been created with 'artificial' boundaries that encompass different ethnic groups, with religious and linguistic cleavages. It has been estimated that only in about a quarter of the states in the world (out of almost 190 states) does over 95 per cent of the population comprise one ethnic group (Nielsson, 1985, pp. 30-1). Thus most states contain potential secessionist nationalisms. In post-colonial Africa attempted secessions and dismemberment of states by intra-state ethnic groups have not been successful. It is not just the fact of religious and linguistic cleavages that such new states must contend with, however, but also the powerful hold of supra-nationalism as in the Arab world and, more particularly, subnational allegiance. In the case of Syria, for instance, 'national' identity and the state have been championed most strongly by a minority whose position would be weakened by expanded boundaries and integration into a pan-Arab space (Drysdale, chapter 6). The Syrian example is particularly instructive since it refers to the rare instance in the map of the world where a state has dissolved and then reformed itself. Elsewhere in the Middle East the right to a firm territorial base lies at the root of many of the present conflicts. The establishment of the State of Israel deterritorialised the Palestinians, who were subsequently dispersed throughout much of the Middle East. In the struggle to maintain themselves in the region, both Israelis and Palestinians employ territorial strategies of occupancy and containment. For the Palestinians, a cultural identity has become a national identity that seeks unity through sovereign control of a bounded territory (Portugali, chapter 10).

Internal unity, based on an accepted nationhood and territorial demarcation, has constituted an important aspect of the practical politics and strategies of nationalist movements. As we have seen, an assertion of a distinctive geography and history is a vital task in the unification of territory claimed by nationalist movements. The latter must draw together a complex set of factors to link the (possibly) diverse people who live in the different parts of the common (national) territory, physically, socially and especially psychologically, so that they identify both with themselves as a 'people' and with the territory. A nationalism thus uses a whole set of symbols to help forge significant national bonds. And, as suggested earlier, a nationalist movement selectively draws from history to legitimate its own hegemonic interpretation of the 'nation'. For national unity to be protected, the enemy or 'other' must be designated outside the (national) society for which nationhood and a state are claimed. At the same time, as Linz (1985) points out, the political implications of this mass support and the control of territory by a sovereign body mean that, in most contemporary societies, nationalism founded on 'primordial' elements (i.e. emphasis on common descent, race, language, cultural tradition, or religion) will have to give way to one more territorially based, so as to appeal and gain support, at least potentially, from the entire population living in the demarcated area irrespective of their origins and traits. It is in this context that we can begin to understand some of the contradictions nationalist movements encounter in unifying what is within and distancing themselves from what is outside.

Nations and States in the World Political Map

Despite the variety of nationalisms and the complexity of economic, social and political forces that engender nationalist movements, nationalism in the contemporary world situates itself in relation to the state. Full self-determination has been rendered meaningless without a state apparatus, for sovereignty is vested in the state. Thus, for example, contemporary claims for the right to decide their own mode of self-determination and type of association with one or more states by indigenous peoples in the 'New World' (including Aborigines in Australia, Indians in the USA, Canada, and Central and South America, Inuit in Canada,and Maoris in New Zealand) face considerable difficulties (Knight, chapter 8; Mercer, 1987). Similar claims for autonomy and territorial reorganisation of a sub-state nationalism, or what has been also termed mini-nationalism or neo-nationalism, are restricted by and dependent upon the existing state which it is challenging.

It is states - acknowledged in international law - which abrogate the legitimate use of violence internally and externally to maintain themselves and impede their dissolution. The nationalism of the state thus reinforces and reproduces the collective sense of identity of its citizens, an identity that attempts to transcend alternative allegiances to territory (localities, regions and supranational) and social affiliations (for example, kin, tribes and classes).

The bureaucratic state (Mann, 1984) demonstrates the penetration of the state into civil society and its means of maintaining power. Mann defines it as being relatively low on what he characterises as despotic power (the state elite is empowered to undertake only a limited range of actions without routine, institutionalised negotiations with groups in civil society), but high on infrastructural power (the capacity to penetrate a wide variety of components of civil society). Capitalist states are bureaucratic states, since they have great power to penetrate and coordinate civil society, but they are themselves mere instruments of that society. As he puts it

> a bureaucracy has a high organizational capacity, yet cannot set its own goals; ... the bureaucratic state is controlled by others, civil society groups, but

their decisions once taken are enforceable through the state's infrastructure (p. 191).

The four tasks undertaken by such states, Mann argues, are the maintenance of internal order, military defence/agression vis-à-vis other states, the maintenance of communications infrastructure, and economic redistribution. These are necessary to civil society and are undertaken most efficiently by a central state. This operates military, economic and ideological power, in a particular way:

> the state elite's resources radiate authoritatively outwards from a centre but stop at defined territorial boundaries. The state is, indeed, a place - both a central place and a unified territorial reach (p. 198).

In terms of Mann's classificatory schema, current socialist states are authoritarian in type - they are high in both bureaucratic and despotic power. To ensure the operation of a complex division of labour and to legitimate it requires penetration of virtually all aspects of civil society; to ensure that plans are followed requires that all power is expressed through the 'authoritative command structure of the state' (Mann, 1984, p. 191). The goal of most capitalist states on the other hand, is to create consensus support for the mode of production via the promotion of ideology, whereas for most socialist states the similar processes lack much allowance for institutionalised opposition to the ideology.

Yet it would be mistaken to assume that there is no place for the nation in the Soviet Union, for example, or that regional support for the regime is simply coerced (G. Smith, chapter 11). Soviet views have evolved over the nationality question and the malleability of national identity (Shevtsov, 1982). Both capitalist and socialist states spatially structure their societies and administer resources to ensure support for the state. The goal of socialist states, like capitalist states, is to create new national identities, and to eliminate much of the basis for backward-looking nationalisms through the achievement of socio-economic equality (Bracht, 1968); some residual regional autonomy may then be allowed which creates no real problem for the state (Levkov, 1980). Achievement of that equality without national differences within states has been rare to date, however.

Analysis of Nationalism

We have so far identified the significance of territory, its delimitation, appropriation and control, for nationalism and the state. There has also been considerable debate about the relationship of nationalism to general processes, such as capitalism and global integration and fragmentation (Ehrenreich, 1983; Stokes, 1986), modernisation and the creation of bureaucracy (Gellner, 1983) and the rise of rational, scientific thought (A.D. Smith, 1971). Questions have been asked, for example, about the role of capitalism in the formation of nation-states. Did the modern form of the nation, and its coupling with the state, emerge in response to earlier processes that had created new forms of civil society and socio-spatial relationships? What is the role of nationalism and the nation-state in the world system? Yet, in these debates, the territorial dimension seems to have been incorporated primarily as a variable with little explicit appraisal of its role in the construction of nationalism.

Within established nation-states, for example, theories of uneven development, internal colonialism and ethnic resurgence generally only implicitly invoke the territorial aspect through the operation of a core-periphery model. More seriously, territorial relations are reified, and social relations in the core and periphery are obscured. At the world scale, Wallerstein simply accepts the difference in scale at which the economy and the polity operate without really explaining it (Stokes, 1986, p. 593). In this schema, strong states were a prerequisite to the rise of nationalism in the core and the periphery of the world capitalist system (Wallerstein, 1974, p. 145; Taylor, 1985).

Surprisingly, territory is hardly mentioned in Gellner's (1983) account of the role of nationalism in the transition from agrarian to industrial society, except in reference to a minimum size at which the modern state must function. Nationalism was necessary in this transition from a complex mosaic of separate, small societies so as to produce a high culture that would bind together all individuals within a state. This was because the reproduction of socialised individuals in a world with a complex division of labour could no longer be satisfactorily carried out by small-scale, social units or localised cultures. Nationalism is a means of imposing cultural homogeneity within the bounds of a given territory; it is thus harnessed by a state undergoing a transition in its tasks.

Gellner's analysis is obviously rooted in the original European state nationalisms, whose experience is the source of much theorising on the formation and integration of the nation-state. Local and regional societies were supposed to lose their territorial attachments, as they began to participate in the modern world of circulation and exchange throughout a unified space that was being shaped in the nineteenth century. However, attention increasingly is being focused today on those peripheral regions of the original nation-states, wherein exist many examples of 'separatist nationalism' and 'internal colonies' (Hechter and Levi, 1979).

The analysis of the diversity of society-state relationships, and the specific configurations in which nationalism emerges as a political response, has recently engendered considerable criticism from geographers. Theorising on nationalism has been insensitive to place and geographical variations, largely because nationalism often is accepted as an autonomous force (Agnew, 1984, 1987). By 'autonomous', Agnew means those approaches where nationalism is considered a thing-in-itself, having causal power, and simply transmitted by social agents. Yet nationalism does not have the same force at all times and places, it waxes and wanes (see also van der Wusten, chapter 12). The geography of nationalism does not just consist of the variations of certain strata, for nationalism is rather a response conditioned by different local environments. Therefore, it must be the political product of a particular set of circumstances, and hence contingent. Whilst Agnew does not deny the impact of global influences, place nevertheless mediates these general processes. Yet, one feels that there is the temptation in the drive to contextualise and elevate a geography of nationalism to its rightful place to try no longer to make sense of the interaction between the general and the particular of local and regional societies in specific historical periods.

It does not necessarily follow that those who criticise autonomous approaches fail to situate nationalist ideologies and movements in relation to the formation of nation-states and the development of a world economy. For McLaughlin (1986), what many theories of nationalism, both structuralist and autonomous, lack is an appreciation of the dialectical relationship of the economic, cultural and political. Many writers commit the 'fallacy of dualism', that is,

> factors that are inextricably linked in the real world are treated as though they do not interact for purposes of theorizing and model-building (p. 306).

Breuilly (1982) has increasingly been cited as a writer who has recognised nationalism as practical politics, and highly varied in space and time, yet he also relates it to the centrality of the state and the general way in which 'capitalism is the major force which both constructs and disrupts 'civil society'. Nationalist ideology can be related to the distinctively modern concern about the relationship between state and society, between public and private' (p. 354). Similarly, from a Marxist perspective, Ehrenreich (1983) tries to develop an explanatory niche between 'a detailed historically specific account and a simplistic, monocausal formula' (p. 10), which could illuminate the 'contradiction between the world-unifying and world-fragmenting tendencies of capitalism ... and nationalism as the ideological offspring of this contradiction' (p. 9).

For McLaughlin (1986), only a dialectical account 'recognizes the existence of deep-seated contradictions, both in relationships of production and in social relationships' (p. 307). He argues that we should look to more historically situated Marxist interpretations in which 'social interaction and class struggles are structured but not determined by the regional setting' (p. 313). It is surprising that a Gramscian analysis has so far not been incorporated in studies of specific nationalist movements (though see McLaughlin and Agnew 1986; and Dulong, 1978), for not only does Gramsci's work develop the 'role of political leadership, organizations, ideologies and superstructural institutions' (McLaughlin, 1986, p. 314), but it also elucidates the formation of class blocs and alliances in particular historical and regional contexts and the role of intellectuals in producing and challenging the hegemony of the ruling classes (see Bennett et al. 1981; for selections from Gramsci's writing; Gramsci 1971, 1977). If we wish to achieve a more contextually and historically rooted understanding of nationalism, we shall have to undertake an analysis that incorporates the relationship between the state and civil society (national and regional), the economic, social and cultural transformation these societies have undergone, the different responses of social classes and groups to these changes, and finally, the internal geographical divisions of the territory claimed by nationalist movements. It is only with a rounded understanding of nationalism that we can appreciate that nationalism is not an inevitable strategy, but rather that it emerges as one amongst several possible territorial and political responses to a changing world.

Where, When and Why?

What the preceding sections have provided is a framework for understanding nationalism. What they do not, and cannot, provide, however, is an explanatory model in the positivist sense; we cannot account for every nationalism by the application of a general formula, and certainly cannot predict nationalist outbreaks, where they will occur, and

when.   The pre-existing cultural mosaic supplies the
resource upon which nationalist movements are based, but
when and where, and for what reasons, such resources will be
drawn upon to promote nationalism and its corresponding
political organisation are determined by individual and
collective actions, as well as the relationship between
state, society and territory.   It is important that we do
not conflate the structural preconditions and triggering
factors associated with nationalism (Orridge and Williams,
1982, pp. 21-2).

        Yet, although it is not our aim to contribute to a
grand theory of nationalism and its relationship to global
processes,   such   as   capitalism,   industrialisation   and
modernisation, we are not, on the other hand, arguing that
each particular nationalism is so distinctive that it can
only be understood as a singular phenomenon, unconnected to
general economic, social and political changes in a world
that is both unified and fragmented.   As we have previously
discussed, the state has become the political organisation
which is recognised as sovereign in the map of the contemporary world, so that concepts and aspirations of self-determination come to be measured by this yardstick.   This does
not mean that in all instances a community or nation,
however defined, necessarily wishes to make the cultural
congruent with the political form of the state.   Many
sub-state groups, including indigenous peoples, may wish to
achieve a more flexible self-determination, but nonetheless
find themselves dependent on the concepts of self-determination practised by the states in which they are located, and
by international organisations.

        In   many   autonomist   and   separatist   nationalisms
(post-colonial and old-state), much effort is devoted to
constructing a nation out of what are in reality disparate
communities and classes.   Nationalism offers an ideological
focus for territorial, social and political unity, where
previously none existed, or had ceased to exist.   Interpretations in terms of internal colonialism and national
liberation have been shared by movements in different parts
of the world, with European movements comparing their
relationship to the centre, and even strategies for liberation, with Third World liberation struggles.   Of course, the
fundamental question to be asked is why different types of
nationalist (sub-state) movements should be seeking to
mobilise a population within a given territory, whose
boundaries are not necessarily fixed.

In Summary

The essays in this book have been assembled to help us
advance the understanding of a major issue in the contempo-

rary map of the world. That understanding requires two linked spheres of activity. The first is the development of a coherent overarching framework, or mode of analysis, that allows us to appreciate the general nature of nationalism and its associated concept of self-determination. The other is the elucidation of particular nationalisms, thereby to increase our awareness of the detailed nature of such social and political movements and their geographical and histori-cal contexts. The two types of work feed off each other; the overall framework enhances the understanding of particu-lars, whereas the latter enables the framework to be more substantially constructed.

All of the essays in this book have been written by geographers. Their discipline has no exclusive claim to the study of nationalism. What the essays demonstrate very clearly, however, is that nationalism is a form of social and political movement firmly rooted in territory, in place and space. Nationalist movements do not just operate territorially, they interpret and appropriate space, place and time, upon which they construct alternative geographies and histories. All too often, writing on nationalism has passed over the significance of territory in nationalist ideologies, politics and strategies.

It has not been our intention to undertake a tour of the world in search of nationalisms, nor to cover comprehen-sively the many interpretations of the nationalist phenome-non. We have, however, attempted to show the complexity and contradictions of nationalist aspirations and strategies in capitalist, socialist and Third World societies.

References

Agnew, J. (1984) 'Place and Political Behaviour: the Geography of Scottish Nationalism', Political Geogra-phy Quarterly 3, pp. 191-206.
Agnew, J.(1987) 'Is There a Geography of Nationalism? The Case of Place and Nationalism in Scotland' in C.H. Williams and E. Kofman (eds.) Community Conflict, Partition and Nationalism. Croom Helm, London.
Anderson, B. (1983) Imagined Communities. Verso, London.
Anderson, J. (1983) 'Nationalism and State Formation' in The State and Society, Social Sciences Course D209, Open University Press, Milton Keynes.
Anderson, J. (1985) 'Regions Against the State' in Changing Britain, Changing World, Social Science Course D205, Open University Press, Milton Keynes.
Anderson, J. (1986) 'Nationalism and Geography' in J. Anderson (ed.) The Rise of the Modern State. Harvester Press, Brighton, pp. 115-42.

Bennett, T., Martin, G., Mercer C. and Woollacott, J. (1981) Culture, Ideology and Social Process. Batsford Academic and Education and Open University Press, London.

Blaut, J. (1986) 'A Theory of Nationalism', Antipode 18, pp. 5-10.

Bracht, H.(1968) Das Selbsbestimmungsrecht der Völker in Osteuropa und China. Wissenschaft un Politik, Köln.

Breuilly, J. (1982) Nationalism and the State. Manchester University Press, Manchester.

Deutsch, K. (1953) 'The Growth of Nations: Some Recurrent Patterns of Political and Social Integration', World Politics 5, pp. 168-95.

Dulong, R. (1978) Les Régions, l'Etat et la Société Locale. PUF, Paris.

Ehrenreich, J. (1983) 'Socialism, Nationalism and Capitalist Development', Review of Radical Political Economy 15, pp. 1-42.

Emerson, E. (1960) From Empire to Nation. Harvard University Press, Cambridge, Mass.

Gellner, E. (1983) Nations and Nationalism. Basil Blackwell, Oxford.

Gottmann, J. (1973) The Significance of Territory. University Press of Virginia, Charlottesville.

Gramsci, A. (1971) Selections from the Prison Notebooks. Lawrence and Wishart, London.

Gramsci, A. (1977) Selections from Political Writings. Lawrence and Wishart, London.

Hechter, M. (1975) Internal Colonialism: the Celtic Fringe in British National Development 1576-1966. Routledge and Kegan Paul, London.

Hechter, M. and Levi, M. (1979) 'The Comparative Analysis of Ethnoregional Movements', Ethnic and Racial Studies vol. 2, no. 2, pp. 260-75.

Juaregui, G. (1987) 'Nationalism in the Basque Country', European Journal of Political Research 16, 587-605.

Knight, D.B. (1982) 'Identity and Territory: Geographical Perspectives on Nationalism and Regionalism', Annals of the Association of American Geographers 72, pp. 514-31.

Knight, D.B. (1984) 'Geographical Perspectives on Self-Determination' in P. Taylor and J. House (eds.) Political Geography: Recent Advances and Future Directions. Croom Helm, London, pp. 168-90.

Knight, D.B. and Davies, M. (1987) Self-Determination: An Interdisciplinary Annotated Bibliography. Garland Press, New York.

Levkov, I. (1980) 'Self-Determination in Soviet Politics' in Y. Alexander and R.A. Friedlander (eds.) Self-Determination: National, Regional, and Global Dimensions. Westview Press, Boulder, Colorado, pp. 133-90.

15

Linz, J. (1985) 'From Primordialism to Nationalism' in E.Tiryakian and R. Rogowski (eds.) New Nationalisms of the Developed West. Allen and Unwin, Winchester MA, pp. 203-53.

McLaughlin, J. (1986) 'The Political Geography of "Nation-Building" and Nationalism in Social Sciences: Structural vs Dialectical Accounts', Political Geography Quarterly vol 5, no. 4, pp. 299-329.

McLaughlin, J. and Agnew, J.A. (1986) 'Hegemony and the Regional Question', Annals of the Association of American Geographers 76, pp. 247-61.

Mann, M. (1984) 'The Autonomous Power of the State', Archives Européennes de Sociologie, 25, pp. 185-213.

Mercer, D. (1987) 'Patterns of Protest: Native Land Rights and Claims in Australia' Political Geography Quarterly 6, pp. 151-170.

Munck, R. (1986) The Difficult Dialogue: Marxism and Nationalism. Zed Books, London.

Nairn, T. (1977) The Break-Up of Britain. New Left Books, London.

Nielsson, G.(1985) 'States and "Nation-Groups" a Global Taxonomy' in E. Tiryakian and R. Rogowski (eds.) New Nationalisms of the Developed West, Allen and Unwin, Winchester MA, pp. 27-56.

Orridge, A.W. and Williams, C.H. (1982) 'Autonomist Nationalism: A Theoretical Framework for Spatial Variations in its Genesis and Development', Political Geography Quarterly 1, pp. 19-39.

Poulantzas, N. (1978) State, Power and Socialism. New Left Books, London.

Sack, R. (1983) 'Human Territoriality: A Theory', Annals of the Association of American Geographers 73, pp. 55-74.

Sathyamurthy, T.V.(1983) Nationalism in the Contemporary World: Political and Sociological Perspectives. Frances Pinter, London.

Shevtsov, V. (1982) The State and Nations in the U.S.S.R. Progress Press, Moscow.

Sivignon, M. (1981) 'A Propos de Géographie Culturelle', L'Espace Géographique 10, pp. 270-4.

Smith, A.D. (1971) Theories of Nationalism. Duckworth, London.

Smith, A.D. (1981) The Ethnic Revival. Cambridge University Press, Cambridge.

Smith, G.E. (1979) 'Political Geography and the Theoretical Study of the East European Nation', Indian Journal of Political Science 40, pp. 59-83.

Smith, G.E. (1987) 'Administering Ethnoregional Stability: the Soviet State and the Nationalities Problem' in C.H. Williams and E. Kofman (eds.) Community Conflict, Partition and Nationalism. Croom Helm London.

Stokes, G. (1986) 'How is Nationalism Related to Capitalism. A Review Article', Comparative Studies in Society and History 28, pp. 591-8.

Taylor, P.J. (1985) Political Geography: World-Economy, Nation-State and Locality. Longman, London and New York.

Tiryakian, E. and Rogowski, R. (eds.) (1985) New Nationalisms of the Developed West. Allen and Unwin, Winchester, MA.

Wallerstein, I. (1974) The Modern World System: Capitalist Agriculture and the Origins of the European World-Economy, 1600-1750. Academic Press, New York.

Weber, E. (1976) Peasants into Frenchmen: The Modernization of Rural France. Stanford University Press, California.

Williams, C.H. (ed.) (1982) National Separatism. University of Wales Press, Cardiff.

Williams, C.H. (1985) 'Conceived in Bondage - Called into Liberty: Reflections on Nationalism', Progress in Human Geography 9, pp. 331-55.

Worsley, P. (1984) The Three Worlds. Weidenfeld and Nicolson, London.

Wright, P. (1985) On Living in an Old Country. The National Past in Contemporary Britain. Verso, London.

Chapter 2

NATIONALIST IDEOLOGY AND TERRITORY

James Anderson

Is it possible to construct a general theory of a phenomenon
as complex and varied as nationalism?
    The most persuasive attempted   theories are either
Marxist or Weberian, but they generally over-emphasise,
respectively, either the economic or the cultural aspects of
nationalism, and typically they cover only some historical
forms of nationalism and not others. Colin Williams (1985,
p. 350) concludes that it is impossible to get 'unity of explanation' in a field which spans 'the development of nation
states since the sixteenth century', the diffusion of
nationalist ideology 'to a universal audience by the
twentieth century', and the recent 'resurgence of autonomist
movements' within some western democracies. In similar
vein, John Agnew (1986, p. 6) suggests that a general theory
is impossible because nationalisms are 'contextually
constituted'.
    Now it is true that nationalisms are to an important
extent constituted by their unique locations in space and
time. This indeed helps to account for their great diversity. Whether expressed in popular sentiments, organised
movements or state policies, they are a territorial form of
ideology. Nations, like states, are not simply located in
geographic space - which is the case with all social
organisations - rather they explicitly claim particular
territories and derive distinctiveness from them. Indeed
nationalists typically over-emphasise the particular
uniqueness of their own territory and history. However, it
does not necessarily follow that general theory is an
impossibility. While fully acknowledging the diversity of
nationalisms and the constitutive influence of their
contexts, we should not overplay their differences and
uniqueness. On the contrary, there are several reasons why
we should retain the goal of general theory and continue to
explore the possibilities of generalising about nationalism.
    Firstly, we need to generalise, not as an alternative
to the detailed study of nationalism in particular regions

and periods, but as complementary to it. Nationalisms
cannot fully be understood simply on a case by case basis.
Despite their diversity they exhibit significant common
characteristics, not least the widespread popular appeal of
nationalism in different situations and its great flexibility in being used by very different social groups and
classes for different and often conflicting purposes. Such
common characteristics, and other general patterns and
trends over space and time, require general explanations.
Secondly, there are sound bases for generalising about
nationalisms. They are constituted by more than their
particular space-time locations and are not reducible to
their immediate contexts. Typically they dwell on their own
past, or rather selective and often fanciful aspects of it,
and they are strongly influenced by other nationalisms past
and present. Moreover, nationalism is a historically
specific phenomenon - albeit one with a long time-span
- which coincides and connects with the rise of capitalism
from feudal and other non-capitalist societies, and with the
development of an integrated world system dominated by
imperialism, itself an outgrowth of nationalism. Particular
nationalisms developed in this same interdependent world,
and not surprisingly they share many of the same problems -
all of which strengthens the case for generalising across
space and time. Thirdly, moving beyond the topic of
nationalism, there is the wider issue of the extent to which
recognition of the constitutive influence of space-time
location sets limits to general social theory. Or, to put
the matter negatively, there is the danger that an over-
emphasis on immediate space-time context and a one-sided
stress on geographical particularity could simply undermine
the possibility of general theory and become an excuse for a
retreat into empiricism.

Thus general theory, in our case a general theory of
nationalism, remains important as a goal. Colin Williams
may be right in suggesting that all the different historical
forms of nationalism cannot be explained by a single theory,
but there is still considerable scope for progress towards
'unity of explanation'. It may be that a unifying theoretical framework which links a number of related theories
dealing with particular forms and aspects of nationalism is
a more realistic goal than a single monolithic theory; but
such a framework would be a significant improvement on the
collection of existing theories some of which employ
inconsistent definitions of nationalism, or pretend to
explain nationalism in general while in fact ignoring or
defining out of existence some of its major phases and
aspects.

One of these aspects which is often under-emphasised
if not ignored is nationalism's relationship to territory,
and this, together with the related territoriality of the

state, provides perhaps the most promising approach to constructing a general theoretical framework. Certainly it is one where geographers might be expected to make a contribution. The approach has several complementary strands. Thus we can study the historical development and spread of nationalism in different historical periods, starting with its origins in the absolutist states of Western Europe and tracing how it evolved with the rise and consolidation of modern states. We can construct typologies based on nationalism's relationships to states and particularly whether it defined the boundaries of the state or simply helped to unify the territory within existing state borders (Anderson, 1986a). We can also critically assess and learn from existing theories, such as those of Ernest Gellner (1964) and Tom Nairn (1977), which have a significant territorial dimension (Anderson, 1986b). And we can analyse the formal structure and general characteristics of nationalism as a territorial ideology. This is the strand of the 'territorial approach' which is developed in this chapter. Drawing on Weberian and Marxist theories, nationalist ideology is related to territory in terms of politics and statehood and the more abstract temporal and spatial structure of the ideology, and then, more concretely, in terms of social classes and economic and cultural development.

Politics and Statehood

The term 'ideology' refers not simply to ideas but to:

> systems of ideas which give distorted and partial accounts of reality, with the objective, and often unintended effect of serving the partial interests of a particular social group or class. Typically they do so by appearing to represent the interests of all the various groups in society (Anderson, 1973, p. 1)

Nationalism is 'territorial' in the sense of claiming specific territory, and it is 'partial' in two senses: internally, the 'national interests' may be more in the interests of some parts of the nation than others, and it may indeed be contrary to the interests of some sections; externally, any particular nation is only a small part of humanity and it faces other nations with differing views.

Nationalism's distortions arise from its partiality, but a central question for any theory of nationalism is how can these distortions and the persistent and widespread belief in them be explained? In idealist interpretations, ideologies may be seen in purely subjective terms, as 'all in the mind', but they persist and have plausibility because they have a partial, albeit often very limited, basis in material reality (Anderson, 1973, p. 2). And so to get a general understanding of how nationalist ideology 'works' we have to investigate its material bases as well as its formal characteristics. These can be briefly listed before being discussed in more detail.

Nationalism in both the Marxist and Weberian traditions (e.g. Lenin, 1970, p. 3; and Weber, 1970, p. 176) is seen primarily in political terms relating to statehood. States also involve territoriality but whereas they are political entities, nations and nationalisms have to be seen as both cultural and political (rather than as one or other, as sometimes happens). They are cultural in the sense that nationalists use particular traditions, folklore or language to weld together a territorial community, a 'nation'; and they are political in that this national community is intimately linked to statehood, either in the shape of an existing state and its economic, military and other policies, or in the sense that the nation (or at least part of it) is seeking to establish a state of its own, something which is often opposed to existing states. Nationalism is thus both internally unifying and externally divisive over space; and it is simultaneously backward-looking and forward-looking over time.

Nations and states are quite distinct entities but they are sometimes confused because the modern, nationalist, ideal - and to a considerable extent, reality - is that they should coincide geographically in nation-states. This comes from two main sources. Existing states, or rather the ruling groups in control of them, wish to appear to speak for a single united nation rather than for sectional or regional interests, in order to achieve legitimacy and popular acceptance from the population of the whole state territory; this is easier if that population has in fact been welded into a single, culturally homogeneous community, and one which sees itself as a nation to whom the state in some sense 'belongs'. Such state-sponsored or state-created nations are sometimes referred to as 'state-nations', to distinguish them from the second main source of the nation-state ideal: nations which have not, or not yet, achieved their own state. For them, statehood, or more access to state power, opens up the possibility that at least some sections or classes in the nation might achieve objectives which would otherwise be unattainable. Achieving a nation state by this second route presupposes the creation of a nation or national movement in the first place, and, typically, the overcoming of opposition from existing states and their state-sponsored nationalisms.

The first route is primarily political in that the state's political boundaries pre-date the formation of a national community; and the state is the main agency in creating or attempting to create a culturally homogeneous nation within pre-existing borders. This was the case with the original 'state nations' such as England and France which emerged from within, and partly in opposition to, absolutist states (see, e.g. Anderson and Hall, 1986), and it has also been the case in former colonial territories. Here the creation of a national community (e.g. out of heterogeneous tribal, language and immigrant groups) is often partly the work of national liberation movements in opposition to the colonial state power, but even in these cases the state provides the initial territorial context.

To the extent that the nation-state ideal is achieved in reality, state boundaries define the national territory. But many states fail to achieve the ideal because the nationalism of the state is opposed by one or more other nationalisms within the state territory. This is the second route to the delimitation of national territory, implying separation from the existing state territory and/or the unification of territories under the jurisdiction of several states. Here the boundaries of the nation pre-date those of the state which is sought, and the absence of appropriate state boundaries means that cultural criteria are particularly important in delimiting the national territory. This is so even where existing states are involved in helping create the national territory, either through assisting the nationalist movement from 'the outside', or through controlling part of the territory being unified, as the Prussian and Piedmont states played key roles in German and Italian unification.

National territories are thus delimited in two main and sometimes directly opposed ways - by existing states, and, in the first instance, by national movements opposed by existing states - and they have to be analysed historically, not least because oppositional nationalisms have often succeeded in becoming state-sponsored nationalisms, and there are other permutations (see Anderson, 1986a). However in both main cases, democratic aspirations and ideals (if not necessarily practice) have been an important ingredient. This is perhaps clearest where oppressed nations claim the democratic right to 'national self-determination' in states of their own. But it also applies to the original 'state nations' which formed within the shell of absolutist states.

Under absolutism sovereignty rested on the hereditary rights of dynastic families and the 'divine right of kings', and the issue of achieving legitimacy in terms of popular acceptance hardly arose. It only began to become an issue with the growth of opposition to absolutism from the rising bourgeoisie and other subordinate classes as in the English and French Revolutions, and particularly in the latter which saw the full development of the nationalist doctrine that sovereignty belonged to 'the people' or 'the nation'. The doctrine implied that the population of an existing state should be culturally homogenised for the effective working of democracy - and of centralised state control, for the converse argument was that where culturally separate groups existed, or persisted, state boundaries should be redrawn so that such groups could exercise their own democracy. Thus if only for self-protection even the most imperialist and autocratic of states have had to take some account of nationalism's democratic implications, notwithstanding that nationalism and nationalists have often been anti-democratic in practice.

In purely political terms we therefore have two powerful bases for nationalism - statehood and democracy. However, to see why they should be associated with national-ism we need to consider its more abstract temporal and spatial characteristics and its relationships with social class, economics and culture.

Time and Space

Nationalism is simultaneously backward-looking and for-ward-looking in the sense that in general a remote past - typically a fabricated heroic version of it - is used to highlight the inadequacies of the recent past and the present, in order to mobilise support for progress and development to a supposedly better future; or, in lower key, present practices are legitimised in terms of a continuity of tradition, whether real or invented. This Janus-faced characteristic of nationalism is a crucial aspect of its flexibility, and its ambivalent mixture of traditionalism and future-promise is central to its potency as an ideology. The mixture can vary to the extent that nationalism can be embraced and used by mutually hostile camps of 'reactionar-ies' and 'progressives' or 'traditionalists' and 'modernis-ers' (e.g. the Ayatollahs and the Shah), all claiming to act on behalf of the nation. Because of such variations nationalisms are often merged with other ideologies which are themselves mutually incompatible, such as racism on the one hand (as in fascism) or variants of socialism on the other (as in some 'Third World' liberation movements). Within any one nation such differences may result in total hostility, but often the variations of 'reactionary' and 'progressive' emphasis represent different strands of currents which can be encompassed within one broad national-ist movement. A nationalism can therefore be seen to offer 'something for everyone', enabling it to draw support from very different sections or classes in the nation.

An important part of nationalism's appeal is the stress on continuity with the past, and this can be used even where it is primarily involved in mobilising for

'modernisation' or 'development'. Nationalism's tradition-
alist face provides reassurance; and the idea of continuity
with a known and familiar past, presented as unique and in-
destructible, is perhaps especially important in situations
of rapid social change. This indeed is the basis for a
psychological theory of nationalism which sees national
identity as a response to the disruptions and uncertainties
of 'modernisation'. However, one does not have to accept
this mono-causal theory to appreciate the importance of
continuity. (Here there is an interesting contrast with
socialism, nationalism's main rival as a development
ideology, for socialism implies much more of a break with
the past. Perhaps notions such as 'African Socialism', or
conjuring up an indigenous communistic past, are attempts to
imply continuity?)

Associations with the past are central to nationalism's territoriality, for territory is the receptacle of the
past in the present. The nation's unique history is
embodied in the nation's unique piece of territory - its
'homeland', the primeval land of its ancestors, older than
any state, the same land which saw its greatest moments,
perhaps its mythical origins. The time has passed but the
space is still there.

Nationalism is two-faced with respect to space as well
as to time. Looking inward, it seeks to unify the nation
and its constituent territory; looking outward, it tends to
divide one nation and territory from another. Leaving aside
for the moment the complications which arise from different
national groups sharing the same territory, nationalism
generally defines people as belonging/not belonging to a
territory and culture, rather than in terms of class or
status divisions, and it seeks to play down internal
divisions and conflicts, partly by externalising the
supposed source of problems. Political objectives are
explicitly couched in terms of the nation rather than a
section of it (e.g. landowners, industrialists, trade
unionists) on the one hand, or in universalist terms (e.g.
'human rights') on the other. Thus in seeking to unify the
nation internally, nationalism simultaneously creates or
exacerbates the external divisions between nations.

Where a national group has an uncontested claim to its
own clearly delimited territory, the 'internal-external'
distinctions for the group translate fairly directly into
'internal-external' in spatial terms. However, while this
is the general thrust of nationalism and the geographical
logic inherent in its ideal of the nation state, in reality
the straightforward equation of group and territory has to
be substantially qualified. In many instances - including
some of the most intractable national conflicts (e.g. in
Ireland and Israel) - the situation is much more complicated: different national groups may have claims on the

same territory; there may be inconsistencies and disagreements about the delimitation of the culturally-defined group
itself; and some of the factors which influence the delimitation of state and, by extension, national territories have
nothing to do with national criteria as such.   Different
national groups may be so intermingled geographically that
it is impossible to speak of 'internal-external' in spatial
terms, except with the implication - sometimes made - that
one or other group should be forcibly removed from the
territory.   The various symbols with which nationalists
choose to delimit their national groups may not coincide
geographically, some with a more limited spatial distribution tending to exclude people and territory which other
symbols claim for the nation; and there is the further
complication that people may have dual national identities
(e.g. Basque and Spanish) as often occurs with regional
nationalism (Anderson, 1985).   Moreover, the non-national
criteria for delimiting territory include such powerful
geographical factors as militarily defensible frontiers; the
'bending' of boundaries to take in valuable natural resources; the creation of 'corridors' to achieve a common
border and unhindered communication with a nearby friendly
power; and the strategic requirements of 'big power'
politics when powerful states impose their own solutions on
smaller nations.

   Thus the actual delimitation of any territory tends in
practice to be something of a compromise between many
conflicting forces; and understanding each particular
'compromise' requires detailed historical analysis of how
nationalism interacted with other factors.   But it also
requires a general understanding of nationalism's own
territoriality, and bearing in mind these qualifications and
complications we can return to discussing our general
framework.

   National territoriality functions like territoriality
in general, but it is a particularly powerful form of
'politico-territorial identity' (Knight, 1982).   Indeed, it
is probably the most powerful in the modern world because of
its Janus-faced temporal characteristics and because of its
associations with statehood and the fact that state sovereignty is now territorially defined.   Taylor (1985, pp.
107-110, following Sack, 1983) notes some of the ways in
which territoriality functions.   It involves control over a
territory - rights to it, access to it, exclusion from it.
More subtly, it reifies power relations, making them appear
more real and 'natural' (and here some extra-social legitimacy may be gained from having 'natural' borders, while of
course nationalists typically depict nations as 'natural'
divisions of humanity).   Territoriality can also displace
attention from the actual relations of power yielded by
dominant groups, as in the neutral sounding 'law of the

land'. More generally territoriality involves a 'fetishism of space' (Anderson, 1973, p. 3) whereby relations between social groups and classes are obscured by being presented as relations between areas or regions or as relationships with areas or regions.

The socially abstract nature of nationalism's temporal and spatial characteristics lend themselves to such 'fetishism'. Belonging/not belonging to a particular territory is a lowest common denominator in social terms, and this helps account for nationalism's flexibility and the fact that it can encompass very diverse objectives. But it would be misleading to see it as simply an 'empty vessel'. In particular, although it has been used by different classes, nationalism should not be considered neutral with respect to class.

Class and Nation

The logic of nationalism is to deny or at least play down the class divisions and contradictions within a nation in order to maximise internal unity. Nevertheless, these fundamental divisions and conflicts of interest exist, and they present major problems for nationalist movements, underlying many of the factional divisions within them. Although the participants may not perceive their interests in class terms - nationalist ideology is often geared to ensuring that they don't - a nationalist movement may be seen as an alliance of different classes, with the 'national interest' being in fact mainly the interest of the dominant class. Invoking the 'national interest' is a means of gaining support from subordinate classes by presenting the partial interests of the dominant class as if they were the interests of all 'the people'. Alternately, if the classes are less unevenly balanced, the nationalist class alliance may be more a battlefield where each class pursues its own interests, giving rise to a set of competing 'national interests', each trying to gain wider support and dominance. Here nationalist ideology may be infused with the demands and rhetoric of class (e.g. the elements of socialism in anti-colonial nationalism and in the recent nationalist revival in Western Europe regions; see Anderson, 1985).

The variations in nationalism stem in large part from the fact that its class composition and support vary greatly over both time and space. Marxists have tended to over-emphasise the role of the bourgeoisie in nationalism, because of the part played by merchant capital in the rise of the original nationalisms, with the unifying of local and regional markets providing a basis for capital accumulation in the first 'state nations' (though even here much of the territorial unification was accomplished by absolutist states and a feudal ruling order for non-capitalist reasons - Anderson and Hall, 1986). Certainly, once nationalism had clearly emerged onto the political stage it was subsequently used by a wide variety of dominant and subordinate classes and groups, exploiters and exploited - including some of the surviving absolutisms (e.g. Tzarist Russia), dissident feudal landlords (Hungary), the gentry (Poland), the petty bourgeoisie, the peasantry, artisans, and professional people in a variety of countries (see, e.g. Hroch, 1985), specific groups such as the clergy (e.g. in the 'Basque Country'), and sections of the working class (e.g. in early twentieth century Ireland where the Marxist James Connolly declared that the cause of Ireland's freedom was the cause of the working class and vice versa). Clearly nationalism cannot be identified simply with the bourgeoisie - even where it dominates the national class alliance, and certainly not where other classes are dominant.

However, it does not follow that nationalism 'functions as a neutral tool or implement' as Jim Blaut (1986) has suggested, nor can nationalism be used equally to further all class interests, for basically two reasons. Firstly, nationalism cannot be neutral in the immediate and obvious sense that a national movement or ideology at any particular time and place has a definite class content, with some classes standing to gain more than others. While this class content can change significantly over time, it generally does so within certain limits because of the importance of historical legacy in nationalism and, especially, the lasting influence of its main formative period(s). This can perhaps be seen most clearly in Lenin's important distinction between the nationalisms of oppressed and oppressor nations, and especially in the way in which the nationalisms of imperial countries such as Britain and France retain strong notions of superiority, including racial superiority (whether in 'benign' paternalistic or in more malignant forms) long after their colonies have gained independence. These imperial nationalisms are inherently conservative and have been shaped overwhelmingly to further and maintain the interests of the dominant ruling classes. In comparison, the nationalisms of formerly oppressed countries and colonies are, on balance, much more likely to contain strong elements of progressive radicalism, though their indigenous ruling groups are not necessarily progressive. Nationalisms in countries which are still oppressed and fighting for political liberation are generally the most amenable to furthering subordinate class interests, because subordinate groups have to be mobilised on the side of the liberation struggle. Yet even in these most favourable cases nationalism is a problematical 'implement', a 'two-edged sword', as far as subordinate classes are concerned. The 'national interest' and the ways in which it is pursued are mainly determined by other more dominant groups, some of whose interests are different if not actually conflicting, though sometimes these differences and conflicts only emerge clearly after the common goal of political independence has been achieved.

The second reason why nationalism cannot be used equally by all classes is perhaps less obvious but more global and fundamental. We cannot assess the class implications of nationalism simply on a case by case basis, important as that is, but have also to see it in terms of a world system of nation- (or would-be nation-) states which are dominated by the capitalist mode of production. Marxists may be guilty of over-identifying nationalism with the bourgeoisie, in that other classes also use nationalism, but the equation can be reformulated in terms of the bourgeoisie now being the dominant class at a world level and the main beneficiary of nationalism not only in the immediate sense that in the majority of nationalisms the bourgeoisie is the dominant force but in the global sense that nationalism underpins the existing world order.

The nationalism of the oppressed and exploited has to be seen in this larger context. On the one hand, national liberation movements can and do succeed in gaining democratic rights and other political freedoms, sometimes by redrawing state boundaries, with subordinate classes such as peasants and workers often gaining significantly from this. On the other hand, national liberation operates primarily at a political as distinct from an economic level and while it often 'revolutionises' state institutions and boundaries it does so within the general 'rules' of the bourgeois order, consistent with the political principles of 'national self-determination' enunciated by John Stuart Mill or Woodrow Wilson, even though individual imperialisms are offended. For national liberation per se is not opposed to class exploitation in general or the capitalist mode of production, rather its targets are domination coming from outside the nation, including perhaps exploitation from outside. But here the enemies are external capitalists not capitalism as such, and the removal of external exploiters may simply clear the way for internal ones.

Thus in economic terms nationalism tends in effect to be at most 'reformist' rather than 'revolutionary', making exploitation more bearable perhaps but not removing it. Indeed, to the extent that reliance on class alliances necessitates the toning down of class demands, even the nationalism of the exploited may confirm the existing class system. Of course this is not inevitable and national liberation movements often include elements of socialism as already noted. Sometimes, however, these are little more than rhetoric to gain working class support, but even where there is genuine working class activity and involvement on an organised class basis (e.g. as by the independent trade unions in South Africa at present) it tends to be politically subordinated to the programme of the nationalist alliance (e.g. the African National Congress). In some cases the socialism becomes thoroughly subverted by nationalism, partly because of being forced to operate within the 'national' arena of a single state (as in Stalin's 'socialism in one country'), and it may become simply an ideological cover for state capitalism in which the working class is effectively excluded from decision-making and exploited by the state as collective capitalist.

Put another way, while subordinate classes - predominantly workers and peasants - often gain from national liberation, they typically gain less as classes (despite perhaps having made a greater contribution to the struggle) than the dominant class. While national liberation may be seen as a precondition for their class liberation, as Connolly declared for Irish workers, it by no means guarantees it and the subordinate may simply stay subordinate and exploited. Thus Connolly foresaw that after the removal of British domination, Irish workers would still have to combat home-grown Irish exploiters. In similar vein Lenin insisted that workers in oppressed nations should fight for national liberation by socialist means through their own independent working class parties (not just independent trades unions as in South Africa today), rather than submerging their class interests in multi-class nationalist parties. That is one measure of how problematical nationalism is for a subordinate class.

To the extent that a subordinate class simply accepts a nationalist frame of reference it disarms itself ideologically and colludes in its own continued subordination, as 'the national interest', defined by dominant classes or groups in control of the state, sets supposedly non-sectional limits to its sectional class demands (and if workers and peasants were to become the dominant classes a nationalist frame of reference would tend to reduce the support they got from subordinate classes in other countries, leaving them to struggle on in a predominantly hostile world dominated by capitalist and state capitalist regimes). Weber, who is perhaps more usually associated with nationalism's cultural basis, recognised the centrality of class interests and class conflict in modern society, and he saw nationalism providing a common consciousness which could transcend that of class within a country. Nationalism could therefore provide a means of controlling class conflict, and particularly opposition from the working class (see Beetham, 1985, pp. 144-147).

Thus, although the logic of nationalism is to play down class divisions, in practice the furthering of class interests, and domination by particular classes within implicit or explicit class alliances, are central features of nationalism. But why is nationalism effective in furthering class interests? Here we have to return to our conception of nationalism as an ideology in the sense that it inflates truths which do have a partial basis in material reality.

Nationalism's ability to unite people with conflicting class interests is geographically-based in the most general sense that people who share the same territory frequently have at least some interests in common simply by virtue of spatial proximity. This degree of commonality may in fact be very partial, but it can be inflated by dominant groups so as to obscure fundamental conflicts of interest. Less abstractly, nationalism has a partial basis in the political fact, already mentioned, that even subordinate groups may achieve real benefits from independent statehood, and it has a basis in spatial aspects of economics and culture.

Economics and Space

In the light of the foregoing discussion, it is appropriate to discuss nationalism's economic basis in terms of capitalism, and here we have to deal with the mutually reinforcing nature of internal and external aspects. Internally nationalism supports the economic and political domination of the capitalists over the rest of the nation (though not necessarily over all of the state, for regionally-based nationalisms can undermine their domination), and externally nationalism strengthens the capitalists' competitive position vis-à-vis other capitals.

This is not simply a political matter of class domination, important though that is. As David Harvey (1982, p. 419) has pointed out, inter-regional class and factional struggles such as those of nationalism have a material basis 'within the circulation process of capital itself'. Some sections of capital are spatially immobile, tied up in particular territories, and the defence of such capital and of the jobs associated with it fosters territorially specific class alliances.

> A portion of the total social capital has to be rendered immobile in order to give the remaining capital greater flexibility of movement. The value of capital, once it is locked into immobile physical and social infrastructures, has to be defended if it is not to be devalued. At the very minimum this means securing the future labour that such investments anticipate by confining the circulation process of the remaining capital within a certain territory over a certain period of time. (Harvey, 1982, pp. 419-420).

The workers involved have an immediate interest in the immobile capital being defended as their livelihoods depend on it, at least in the short term.

This economic basis for territoriality and class alliances applies at other spatial scales beside the national, but clearly it is most important at the level of the nation-state (or potential state) territory, for this is the main level at which flows of labour are politically controlled. Thus the economic basis of nationalism involves much more than the creation of the larger and more unified internal markets which were important in the rise of merchant capitalism. At least since the arrival of industrial capitalism, labour power has become the central commodity within the 'internal markets', and it remains much less 'internationalised' than other commodities for reasons to do with political control, culture, and the relative spatial immobility of labour compared to other factors of production (which helps explain the persistence of nationalism despite the extensive internationalisation not only of markets but also of production in other commodities).

Just as some sections of capital are more involved in spatially immobile investment than others, so some sections of labour have more to gain than others from a territorially-based alliance with capital. Here we have an important material basis for the 'fetishism of space', of which nationalism is a prime example, for there really is competition on an 'area versus area' basis as

> each alliance seeks to capture and contain the benefits to be had from flows of capital and labour power through territories under their effective control (Harvey, 1982, p. 420).

A clear example of this is protectionist policies which bring immediate benefits to some workers (e.g. in the form of jobs or higher wages) as well as to capitalists (e.g. in terms of orders and profits) - and Tom Nairn (1977 and 1983) has inflated a general theory of nationalism out of such motivations, seeing it (rather one-sidedly) as a response from relatively backward areas to competition from the more advanced (see Anderson, 1986b).

Typically, though not exclusively, it is relatively privileged sections of labour which form such class alliances, their relative privileges over other workers forming the essential cement for the alliance - for example, the marginal privileges of Protestant workers in Northern Ireland who from the late nineteenth century espoused a particularly virulent form of British nationalism in alliance with their Unionist employers and in opposition to Catholic workers. Such relative privileges arise out of the processes of production, but they are also consciously fostered and amplified as a mechanism of political control. Although, being divisive, they can be clearly seen as contrary to workers' long term interests as a class, they are often more immediately tangible to individual workers or sections of workers than the long term interests of the class as a whole.

Perhaps the clearest case of the use of immediate interests for political control is contained in Lenin's conception of a privileged 'labour aristocracy', and the policy of 'social imperialism' propagated in imperial countries such as Britain and Germany around the turn of the century. It shows how the internal and external aspects of nationalism can be mutually reinforcing to the overall gain of dominant classes. The internal counterpart of an aggressively imperialist foreign policy was the ability to pay higher wages and, especially in Germany, a strategy of state welfare provisions and other social reforms to benefit the working class, and especially the more 'respectable' sections of it, to undercut the growing challenge from socialism and working class internationalism. The internal reforms of 'social imperialism' were made economically possible by imperialism, and they gave important sections of the working class a stake in imperialism and colonialism, binding them materially and ideologically to their respective states and ruling classes. Thus these policies simultaneously reduced the chances of workers making common cause with workers and other subordinate classes in other countries and especially in the colonies (their success being shown in 1914, and since, when workers from different countries have nationalistically marched off to kill each other).

Territorially-based class alliances do not remove class antagonism but they do dampen down class conflicts and make it easier for ruling classes to contain them. As we saw, nationalism tends to be at most 'reformist' rather than 'revolutionary' in economic terms, as far as the working class is concerned, and, conversely, its 'reformism' is generally nationalistic in perspective. On the other hand, while underlying class antagonisms make territorial class alliances inherently unstable, these alliances can gain some permanence if they are bolstered by nationalist ideology and institutionalised in nationalist organisations and, especially, state policies. To put it another way, while the material basis of nationalist class alliances may be to an important extent economic, economic benefits and privileges are rarely if ever a sufficient basis for lasting alliances. These are also held together by cultural and political ties in the case of 'state-nations', and by largely cultural ties in 'non-state' nations which are still seeking statehood.

Culture and Community

Culture is indeed so important a basis for nationalism that the concept of nation is sometimes mistakenly seen as purely a matter of culture (in Europe, most commonly of language). It is possible, for instance, to find 'cultural nationalists' who subjectively are apolitical and see no connection with statehood, never mind something as grubby as economic interests, though preserving their culture invariably implies something about state policies, if not necessarily independent statehood, and economic self-interest is rarely far away from even the most spiritual of nationalist movements. The importance, and the complementary nature, of nationalism's cultural aspect can however be clearly seen through contrasts with its political and economic aspects.

In an interesting recent theory of nationalism which centres on culture, Benedict Anderson (1983) explains its origins in terms of a switch in the focus of non-local identities from the large multilingual cultural entities of the 'world' religions (e.g. Islam, Buddhism, Christianity, etc.), to the unilingual nation or nation-state. This switch in consciousness arose initially in early modern Europe through the interaction between the rise of capitalism, the development of vernacular languages out of more localised dialects, and the spread of printing. Printing enabled the mass production of books and it removed the priestly monopoly over written communication which had been mainly in Latin, thereby fostering the establishment and use of written vernaculars, at first in absolutist states. The national consciousness which resulted from such 'book-bound' cultures was thus focused on territorially-based communities connected to statehood - non-local in character but without the pretensions to universality of the 'world' religions - and they provided the prototypes for later nations all of which base themselves on a limited territory and in some sense a 'vernacular culture', if not always on a vernacular language.

Indeed it is in connecting culture with politics and statehood that nationalism shows its modernity. John Breuilly (1982, p. 5) makes a telling point in noting that Dante advocated literary Italian as the language for an Italian nation, and he advocated ideal forms of government, but there was no connection between his cultural and political concerns - nations did not feature in his medieval political schemes.

Benedict Anderson refers to nations as 'imagined communities', but national solidarity and national class alliances (whether or not cemented by privileges) are in fact now easier to 'imagine' than international class solidarity - which is a measure of the contemporary political importance of territory and culture. National and other forms of territorial solidarity are by definition more spatially limited, and more immediate and 'parochial', more based on day-to-day experience and 'common sense', and less reliant on a theoretical understanding of how society functions, less reliant on knowledge and contacts far afield. They are sustainable on a more partial knowledge of reality - are, in a word, more ideological - and clearly they have less problems of communication than international solidarity which crosses cultural boundaries. There are important cultural differences between different classes in the same territory but they are generally less than the differences between people of the same class in different territories, especially where mutually incomprehensible languages are spoken.

In Ernest Gellner's (1964, 1983) Weberian theory of nationalism (which has some similarities to Anderson's), culture is communication, and its significance for national solidarity stems from the fact that in modern society where many social encounters are non-local, impersonal, and ephemeral in character, people have to share a means of communication and hence a common identity which is also non-local, non-sectional and non-class. A shared cultural identity, an ability to 'speak the same language', is central to the inter-class solidarity of nationalism.

Conversely, the absence of a shared culture can give rise to a sharper form of class conflict, and to a more or less direct coincidence between class and national struggles, as in nineteenth century rural Ireland where the Irish peasantry for the most part faced an essentially British landed class, though such class-nation coincidences are much less common than the inflated theory of 'internal colonialism' suggests (see Anderson, 1985). Even the most oppressed nations generally have their own exploiting classes or sections of the population who act as agents for foreign exploiters. However, the class structures of oppressed nations are often significantly different from those of the dominant or oppressor nations. For instance, in many of the smaller 'non-state' nations of Europe, the bourgeoisie and the remnants of feudal ruling classes, to the extent they existed, were incorporated or integrated into the dominant culture of the national community of the state, the 'state nation' (e.g. the Basque financial-industrial oligarchy in Spain). This absence of their own national bourgeoisie and own state lay behind Marx's and Engels' notion of 'history-less nations'. This notion, an Hegelian 'hangover', reflected their identification of nationalism with the bourgeoisie which was then progressive in breaking down feudal and other archaic barriers to development; but the notion failed to recognise that other groups could forge national movements which in some cases would achieve statehood, or at least a continuing 'place in history'.

34

A Czech Marxist historian Miroslav Hroch (1985) has produced a very interesting analysis of some of these smaller European nations, including the Estonians, Lithuanians, Slovaks, and Norwegians. He shows the importance of variations in class structure and culture, and, most originally, how these variations gave rise to geographical patterns in the way nationalist movements developed. For example, in spatial terms the initial nationalist movements generally developed from fairly compact regions which were intermediate in terms of social change and incorporation into the dominant national culture, rather than in remote and traditional peasant regions, on the one hand, or in the major centres of population and industry, on the other. The 'core' areas of early nationalist agitation tended to be ones which were dominated by small scale artisan production for local markets - as distinct from industrial manufacturing - and the more fertile agricultural regions producing for local and distant markets - as distinct from farming regions with a large subsistence element. In corresponding social terms, the activists of nationalism were also intermediate, being drawn mainly from the petty bourgeoisie, professionals, clerics and other 'middle ranks' of society, rather than from the poorer peasants and urban workers who generally had other concerns and lacked the necessary resources, or from the larger merchants and industrialists who were mostly integrated into the dominant national culture.

Marxists have traditionally tended to under-estimate the importance of culture to nationalism (as Weberians often underplay its economic and class aspects), and partly because of this they have also tended to under-estimate nationalism's emotional and political power. Among the more significant exceptions was the Austrian Marxist Otto Bauer (see e.g. Bottomore and Goode, 1978, pp. 102-117; and Burghardt's chapter in this book), though his influence was lessened by political differences with Lenin whose largely strategic and political approach to nationalism was more democratic in its insistence on national rights to self-determination (see e.g. Lenin, 1970). Bauer wished to preserve the multi-national Austro-Hungarian state and he was mistaken in trying to divorce national culture from politics. However, because of his concern with culture, he had a better understanding of national consciousness than most Marxists. It is significant that he, like Max Weber, drew on Tonnies' distinction between Gemeinschaft and Gesellschaft types of social grouping, for the distinction helps to explain nationalism's emotional appeal and the contrasts, and possible complementarity, between the national cultural community and 'purely' political forms of organisation.

A _Gemeinschaft_ is a community based on a feeling of the members that they belong together as a distinct group, with a subjectively held common identity and sentiment of solidarity, as, for example, in the communities of traditional societies. A _Gesellschaft_, by contrast, is an impersonal association formed consciously for specific purposes. More typical in modern societies, examples include trade unions and political parties. For Weber, and for Bauer, nations belonged to the first category, in which subjective feelings of belonging were of central importance, even though rooted in various objective factors of culture, tradition and territory from which the feelings of distinctiveness and uniqueness were derived. While it might perhaps be more accurate to see national movements as containing a substantial element of _Gesellschaft_, in that they have conscious political purposes (and Weber, as already mentioned, saw nations as linked to statehood), nevertheless the national community, with its basis in tradition and in pre-existing cultural differentiation, at the very least gives a strong illusion of _Gemeinschaft_.

The state, on the other hand, is clearly a _Gesellschaft_ in Weber's terms, and it needs to harness the freely given allegiance of the _Gemeinschaft_ national community in support of its power; while the nation, reciprocally, needs the power of a state to preserve its distinctiveness. But while the national community can complement the purely political institution of the state, largely because both are territorial, it is inherently much less supportive of clearly sectional _Gesellschaft_ political organisations such as trade unions. Indeed for Weber, writing from a bourgeois standpoint (Beetham, 1985, pp. 144-7), nationalism supported the state largely because it could transcend and undermine class consciousness in the lower orders. Part of its appeal, as Weber was aware, was that the nation is a type of status group where, unusually, status 'superiority' is available to the lower orders in the national community if its nationalism claims superiority over other nations (Beetham, 1985, p. 122); and, as we saw in the case of 'social imperialism', it can offer economic as well as psychological rewards. Racism similarly can offer status 'superiority' to the bottom as well as the top of an ethnically defined community and, not surprisingly, ideologies of national and racial superiority sometimes go hand in hand (see e.g. Miles, 1985).

Some Conclusions

The boundaries of national territory may be staked out with cultural 'markers' but just as economics is not a sufficient basis for a lasting class alliance, so culture on its own is not a sufficient basis for territorial delimitation. Many pre-nationalist cultural entities (including some languages) have disappeared or been killed off. Whether or not they are made the basis for a national movement - and one that achieves some success in at least creating a territorial community if not an independent state - depends on economic factors and, above all, on the outcome of political struggles involving states and nations. A general theoretical framework has to integrate all these elements and cannot afford to be one-sidedly 'cultural' or 'economic' or 'political'. It has to encompass nationalism's links with statehood and democratic ideals, its two-faced nature with respect both to time and space, and its use to further different class interests whether of domination or liberation, as well as its basis in spatial aspects of economics and culture.

Particular nationalisms require detailed empirical analysis but this analysis has to be informed by a general theoretical understanding of how and why nationalism arose and how it 'works' as an ideology. We have to generalise in order to make sense of nationalism's great diversity and flexibility. We can do this without denying its diversity, though a general theoretical framework linking a number or related theories in a consistent fashion is probably a more sensible goal than a single comprehensive theory. Territoriality provides a key, and a central focus for the 'territorial approach' is the formal structure of nationalism as a territorial ideology.

Acknowledgement

The chapter benefited from helpful comments on earlier versions: at the IGU Conference on nationalism in San Sebastian; from Allan Cochrane and Doreen Massey at the Open University; and from staff and graduate students at seminars in the Geography Departments of Cambridge University and the London School of Economics.

References

Agnew, J. (1986) 'Nationalism: Autonomous Force or Practical Politics', paper presented to the Institute of British Geographers Annual Meeting, University of Reading.

Anderson, B. (1983) Imagined Communities: Reflections on the Origin and Spread of Nationalism, Verso, London.

Anderson, J. (1973) Ideology in Geography, Antipode, 5(3), 1-6.

Anderson, J. (1985) Regions against the State, Unit 26, D205, 'Changing Britain, changing world: geographical perspectives', The Open University Press, Milton Keynes.

Anderson, J. (1986a) Nationalism and Geography, in Anderson, J. (ed.) The Rise of the Modern State, Harvester Press, Brighton, pp. 115-42.

Anderson, J. (1986b) On Theories of Nationalism and the Size of States, Antipode, 18, 218-232.

Anderson, J. and Hall, S. (1986) Absolutism and other Ancestors, in Anderson, J. (ed.) The Rise of the Modern State, Harvester Press, Brighton, 21-40.

Beetham, D. (1985) Max Weber and the Theory of Modern Politics, Polity Press, Cambridge.

Blaut, J.M. (1986) A Theory of Nationalism, Antipode, 18, 5-10.

Bottomore, T. and Goode, P. (1978) Austro-Marxism: Texts Translated and Edited, Clarendon Press, Oxford.

Breuilly, J. (1982) Nationalism and the State, Manchester University Press, Manchester.

Gellner, E. (1964) Nationalism, in Thought and Change, Weidenfeld & Nicholson, London.

Gellner, E. (1983) Nations and Nationalism, Basil Blackwell, Oxford.

Harvey, D. (1982) The Limits to Capital, Basil Blackwell, Oxford.

Hroch, M. (1985) Social Preconditions of National Revival in Europe: A comparative analysis of the social composition of patriotic groups among the smaller European nations, Cambridge University Press, Cambridge.

Knight, D.B. (1982) Identity and Territory: Geographical Perspectives on Nationalism and Regionalism, Annals of the Association of American Geographers, 72, 514-531.

Lenin, V.I. (1970) The Socialist Revolution and the Right of Nations to Self-Determination, in Lenin on the National and Colonial Questions, Foreign Language Press, Peking.

Miles, R. (1985) 'Recent Marxist Theories of Nationalism and the Issue of Racism', paper presented to International Sociological Association conference on Marxist Perspectives on Ethnicity and Nationalism, University of Belgrade.

Nairn, T. (1977) The Break-up of Britain: Crisis and Neo-Nationalism, New Left Books, London. Edited extracts: Nationalism and the Uneven Geography of Development, in Section 2 of Held, D. et al. (eds.) (1983) States and Societies, Martin Robertson, Oxford, in association with the Open University, 195-206.

Sack, R.D. (1983) Human territoriality: a theory, Annals of the Association of American Geographers, 73, 55-74.

Nationalist Ideology and Territory

Taylor, P. (1985) <u>Political Geography: World-Economy, Nation-State and Locality</u>, Longman, London.
Weber, J. (1970) <u>From Max Weber: Essays in Sociology</u>, ed. H.H. Gerth and C. Wright Mills, Routledge & Kegan Paul, London.
Williams, C.H. (1985) Conceived in bondage - called into liberty: reflections on nationalism, <u>Progress in Human Geography</u>, 4, 331-355.

Chapter 3

NATION-STATE BUILDING IN A "NEWLY INDUSTRIALIZED COUNTRY":
REFLECTIONS ON THE BRAZILIAN AMAZONIA CASE[1]

Bertha K. Becker

Incorporated into the world economy as a colony and a huge
resource frontier, Brazil has achieved the position of a
'newly industrialized country', or a part of the 'semi-
periphery' in the last two decades. The process of change,
initiated after the 1929 world crisis, developed in the
1950s and accelerated after the 1964 military coup. The
state experienced national growth based on an association
between nationalism and state intervention which used space
as a privileged instrument. The process proceeded varyingly
in different regions. This chapter demonstrates the process
of integration of Brazilian Amazonia as part of nation-state
building.

Within this context appropriation and control of the
Amazonian region became strategic. Occupying more than half
of the country's territory, it is one of the largest and
richest frontiers in the world. Powerful strategies
developed by the authoritarian state in the last twenty
years seek regional integration, which contributes to
economic growth, ideological control and social transformation at national level. At the regional and local levels,
however, intense social conflicts take place, due to the
contradictions between this form of integration on the one
hand and the political failure to rule and control its cruel
effects on the other. The debate over Amazonia attained a
symbolic meaning, dramatizing the nation's encounter with
its destiny. Under the new circumstances brought about with
the end of the military regime in 1985, regional integration
poses the problem of how to achieve a democratic territorial
organization. Since planning is not a matter of a single
actor's decision any longer, it expresses the new form of
state management under the pressure of various social and
political demands.

In the first section of the paper, some background
elements are presented. The second section deals with the
state strategy for regional integration in the 1970s. The
third and fourth sections focus on the new trends in

nation-state building in the 1980s - global strategies through multinationalization of a state enterprise, and rise of specific local powers. Finally, present trends are discussed.

Notes on Nationalism, State and Space in Brazil

Nation-state building in Brazil is based on a close association between nationalism and state intervention. On the one hand, nationalism has been a basic factor for the expansion of the public sector, developed because of: the insufficiency of private initiative in sectors considered as strategic; the rejection of economic colonization by foreign enterprise, which is in the core of national cultural identity; the military ideology that foresees Brazil as a continental power; and the reinforcement of the state structure against regional interests. After the 1964 coup, when the 'National Security Doctrine' became central in the developing military ideology, security and development became mutually sustaining. According to this doctrine the struggle to survive in a world of conflict and war requires from states the definition of permanent national objectives and maximization of economic growth at any price; strong (military) security is essential to preserve democracy from subversion and attain rapid economic growth. Since then, national security's domain has extended to almost all national activities. The state acted as a mediator between national and international industrial-financial interests and its own conception of nationalism; a tripod model, based on the state, international and local capital, became the basis of economic growth. Development imperatives - national-scale problems, use of the resource potential, reduction of the economic gap - and political imperatives referring to the need to face international protectionism and pressures, led to the expansion of the public sector and to the process of multinationalization of some state companies.

On the other hand, as instruments of industrialization, resource development or international politics, state-owned enterprises are the major means for national development. They represent a third of the gross fixed capital formation each year and are responsible, to a great extent, for the country's position as newly industrialized, at the same time strengthening state power and becoming a basic condition for nation-state building. Energy, transportation, mining, part of the steel industry and credit are appropriated by the nation through the state. State intervention, therefore, has been playing a fundamental role in nation-building although with a poor performance in social and political terms. Benefits of economic growth are

41

appropriated by the few. Political institutions and a real civil society are still amiss: the irrationality of the state management and the power of corporations - private and public - grow.

The formation of the national state is closely bound to the imposition of a new spatial order by the state. Since colonial times and after political independence, the expansion of the internal frontier sustained the unstable reality of an economy dependent on the international market. After 1929, when industrialization developed, the state-space relation became closer, particularly with the dictatorial government in power after 1964, which, according to national security doctrine, needed to control economically and politically all the territory. From then on state action aimed to remove material and political barriers to modern capitalist expansion and to the centralization of power. According to national priorities, it created the conditions to eliminate the economic and political 'islands' on which Brazilian economy and society were organized, thus integrating the space economy where regions acquired national functions.

'Westward bound' in the 'thirties and 'forties; 'Energy and Transportation' in the 'fifties; 'Regional Development' in the 'sixties; 'National Integration' in the 'seventies: these discourses were to galvanize the nation, being in fact strategies to the quick modernization of the territory and for centralizing state power. Regional development policies were a basic strategic tool in this process, since they expressed the concentration of efforts on selected areas. The first step in the regional policy was the institutionalization of macro-regions through the creation of regional agencies managed by federal government - Sudene for the northeast region (1959), Sudam for Amazonia (1966), Sudeco and Sudesul for the Centre-West and South, respectively (1967). These were political acts aimed at neutralizing old regional oligarchies and organizing the basis for the modern space appropriation.

Between 1974-79, growth pole policy expressed the shift of the state's strategy from the regional to the subregional level, selecting within the underdeveloped regions a great number of areas for their comparative advantages according to national priorities. In the 1980s, development programmes are being formulated for more specific purposes and involve huge areas (mining, river basins). National urban policy was a key element in the process of regionalization implemented by the federal government, complemented by transportation and agricultural modernization (via credits, subsidies and incentives). In this way, a frontier 'know-how' was generated, relating to the extension of all kinds of circulation networks, subsidized build-up of entrepreneurs and creation of a substratum of a mobile labour force in order to meet the demand created by the new spatial order. The process of territorial integration was identified with the process of nation-state building since a deep transformation of the society occurred parallel to a growing role and power of the state: the institutional frame was organized and the institutional network extended through the territory; state-owned enterprises were created; new social actors were brought into the scene (such as the technocracy, the urban middle class) to support the new alliance constituted by the tripod model. As a result, the national space became urbanized under the sign of growing metropolization and Amazonia was thought of as the great national frontier, whose appropriation and control became strategic for the formation of the national state, in economic ideologic and political terms.

National Scale and Ideology: the State and the New Meaning of the Frontier

The new meaning for the frontier arose as part of the reflection about the roles of state and space as well as about the relation between them. After the Second World War theoretical thinking recognized that daily reality, and not the general economic condition alone, forms the basis for the reproduction of the social production relationships, and that society as a whole, and therefore space as a whole, becomes a condition for the general reproduction. In this context the primary role of the state changes: it is to legitimize the reproduction of wide domination relationships controlling all spheres of the social life through the production of a social space involving a hierarchic set of locations, laws, conventions and institutions which are the state itself, structuring the social division of labour and obtaining the passive acceptance of citizens.

Resources, techniques and conceptual ability allow the state to deal with space as a whole, in large scale, not only controlling the production units but also creating a double-control net, technical as well as political. It then imposes a spatial order linked to a global space praxis and conception (and thus rational, homogeneous and logistic) focused on common interest, opposed to a local space praxis and conception, focused on the 'space production agents', private interests and personal goals: 'It thus engenders, not a new space, but a specific product of the private-public cleavage, expressed in a double-character global/fragmented state' (Lefebvre, 1978). In turn, protest movements of space-users break out, claiming not only for working conditions, but also for daily life, for the whole space.

When producing its political space, as a mediator between the economic reality of the global scale and living experience in the local scale, the state gains relative autonomy, which sets up a contradiction between its accumulation and its legitimation functions.

It is within this context that the frontier acquires a new meaning. Frontier is not a synonym for free land open to pioneers for economic appropriation. It is the social space, and not the physical one alone, that distinguishes the frontier. The alternative hypothesis is that the <u>frontier consists of non-fully-structured space, a space in process of incorporation to the global/fragmented space</u>, with unfinished capitalist forms of production and social relations, and therefore with a high political potential (Becker, 1985a). In other words, the frontier is the space where virtually all actors involved have an expectation of wider reproduction, but this reproduction is still uncertain.

In the case of Amazonia, the political potentiality contained in such a wide geographic scale makes it the strategic space par excellence for the state, which strives to control and structure it the fastest, in order to integrate it in the global space, manipulating, at the same time, the image of an alternative space. State interests, therefore, go beyond the economic ones of international and national investors. For the nation, the development path taken by the frontier becomes a symbol and political factor of major magnitude. It is in Amazonia, therefore, that the whole set of state spatial interventions are clear, since they may be observed from the moment they are being created.

State strategy for a quick territorial and social integration of the frontier in the 'seventies is contained in implicit and explicit policies, which create the necessary conditions for this integration, mainly: 1) networks, 2) subsidies for land appropriation, 3) labour mobility.[2]

(1) <u>Space integration and ordaining networks</u>
These are a condition for the economic and political integration of the frontier; among them the role of the urban network is highlighted.

The pioneer road network was built, first, to ensure the articulation of the territory. The Belem-Brasilia road, a north-south axis, was built the end of the 'fifties; in the 'seventies, the region was cut in all directions by a network of federal roads. Along some of these axes, an urban network also was established to support official colonization projects. Then came the highly efficient telecommunication network that was intended to support 'psycho-social' integration. In the 'eighties, another kind of network was set up, the hydroelectric one, aimed at providing basic support for aluminium production in large projects.

Nevertheless, it is the urban network that plays the role of logistic basis for the project of quick occupation and ordaining of territorial and social space in the frontier. The state project for the occupation of Amazonian frontier involved urbanization as a previously-set deliberate strategy, and was viewed as a means for fostering regional economic development (Racionero, 1978).[3] Such strategy is conveyed in the 'rural urbanism' present in private and official colonization projects along federal roads (1970-74), as well as in the 1975 policy of selective growth poles, Polamazonia.

Two primary roles of urban centres in frontier integration underlie the enunciation of strategy by the state. First, they promote the physical integration and ordaining of the territory, as a basis for general circulation, and above all as a basis for labour market organization, since they are the 'locus' where change in social production relationships takes place, where the concentration and reallocation of work force (which should be available, mobile, but gathered in selected points within the territory) happens. These elements give them a dominant character of <u>reproduction spaces</u> (Becker, 1985d). The smaller the centres, the more specific is their function as a circulation basis for manpower, the more precarious is equipment and shorter is their lifetime. Second, they promote social space integration, as a locus for political ideological state action, for the Church and for hegemonic groups of the non-monopolistic fraction forming in the new local society, and as the place where the resocialization of the work force for its role in society is carried out. They also sustain the ideological image of the frontier as an alternative space, since they offer possibilities of obtaining land or a job.

In these ways, frontier expansion is carried out in a context of quick urbanization. The urban population grew between 1970 and 1980 from 1,652,688 to 2,720,140, from 36 to 43 per cent of total population in the region. In Amazonian urban centres as a whole, true urban 'circuits' can be distinguished, each one relatively independent and at a different scale: (a) urbanization as a result of state initiative, which appears under two shapes: guided urbanization, happening in private and government colonization areas, and urbanization in sub-spontaneous settlement areas, resulting from the inducing action of the state; (b) urbanization linked to work force circulation - hamlets and villages which make up the informal circuit within the formal urban network; and (c) urban spearheads of great projects belonging to transnational corporations - 'company towns' that are self-supporting and linked to resource exploitation.

(2) Subsidies for monopolist land appropriation

Land concentration in the hands of the state, and its controlled distribution, is a primary condition for state power. Land located along federal highways (100km on each side) and along international borders is, by decree, declared to belong to the federal government. The creation of new territories under central government jurisdiction (to solve conflicts over land or for development programmes) by overlapping the official administrative territorial subdivision, and by land title regularization, were the main mechanisms utilized to transfer land to the hands of central government.

Land monopoly, or private land appropriation, is a condition to generate future rent either through its productive exploitation or through financial speculation, a means to obtain subsidies and credit, and it is also a condition for the reproduction of dominant ruling classes. A tax incentive mechanism redirected investments by transnational and national firms - located in the Southeast region or abroad - to the frontier. Large tracts of land were appropriated, however, with low productive investment. The state promoted the nationalization of the land as well, through the distribution carried out by official colonization projects (Velho, 1985) in strategic locations, thus inducing the formation of small capitalized producers.

(3) Labour mobility

This is the main condition for the expansion of capitalist forms of production on the frontier. The formation of the regional labour market does not happen according to the classical model of full proletarization, but with labour mobility (Gaudemar, 1976): labour is constituted by wage-earners (permanent and temporary) and small producers that eventually sell their work force, engaging in assorted tasks within rural and urban activities. Labour mobility is a process within which takes place simultaneously the social differentiation of peasants; it is a spatialized process of social fragmentation. The formation of the regional labour market implies a process of migration and mobility basically induced by the state. Through implicit and explicit policies, the state promoted the national mobilization of migrants to the frontier. Advertisements, employment in public works, access to land, credit, and urban facilities were powerful tools to attract migrants. Since employment is temporary and access to land is limited and also temporary, a vast manpower reserve was created in the region, able to perform various tasks and also to produce foodstuffs for their own consumption.

This form of mobile labour does not seem to be a transition for full proletarization, having come into existence as a way to bypass the labour-capital contradiction in areas of 'unfinished' capitalism like Brazil, and maybe even Latin America as a whole. Its intensity is much higher in Amazonia, due to the relatively unstructured character of the region.

The New Reality of International Scale: High Technology, Global Strategies and the Frontier

From the 1970s on, there came into existence a new frontier for the twenty-first century, with the technological revolution in the fields of electronics and communication, which, by creating a new form of production and social organization based upon information and knowledge - high technology - reorganized the basis of the accumulation model (Castells, 1985).

Transnational corporations and international financial agencies reinforce a global economy whose interdependence is virtually complete, extending industrial production to developing countries.[4] Work force segmentation, engendered by the requalification/disqualification polarization when faced with high-technology activities and production segmentation, combining resources on a world scale, assure globalization. The implantation of a new planetary order redefines the role of national states. Countries, in the face of expansion of corporations, are not the economic unities of historical reality any longer; the state loses control over local decisions of corporations, over the productive process as a whole, and over its space, facts aggravated in developing countries by foreign debt (Becker, 1985c).

Globalization seems, however, to be much more complex than it looks, since a process of multinationalization of public enterprises is occurring, particularly in developing countries. Therefore transnationalization is also a contemporary way for nation-state building. To assure the country's economic growth and to face foreign corporations, the state enters in the global order, providing conditions for state-owned enterprises to produce a transnational space, and at the same time assuring its own strength. Through this strategy industrialization develops and some of the developing countries change their role in the world-economy, becoming 'newly-industrialized' or semi-peripheries.

Although global strategies concern Brazil and the world as a whole, Amazonia, as the major energy reserve in the planet, becomes a geopolitically privileged space for supranational corporation action, both private and public.

Due to the hugeness of the territory, to the riches that it contains, and the non-existence of social organizations able to resist the new appropriation, Amazonia is a space where it is feasible to implant new structures, to exert a monopoly upon means of production (raw materials, cheap manpower and land), a space where it is easy to open markets for high technology, and to extend the control of world financial markets.

Corporate interests are in the monopolist appropriation of space and resources as a reserve, and, in the short run, the implantation of mining projects related to mineral exploitation at a world scale. The exploitation of bauxite and iron sets up a new state in the frontier, the large-size mining industry state, aimed at exports. State interests surpass economic interests, as indicated above. Exploitation of regional resources under (at least partial) national control is vital, and contributes to the expansion of state-owned enterprises and eventually to their multinationalization, as has been happening to Companhia Vale do Rio Doce (CVRD), the Brazilian state enterprise which acts in mining.

Among thirty-three large industrial or infrastructural projects (each one with one billion dollars investment or more) that materialized as part of government strategy for the 1980s, six are located in Amazonia. Two of them are controlled exclusively by state enterprises - the Iron Carajas Project which integrates mining, railway and harbour schemes to export iron ore controlled by the CVRD, and the Tucurui Hydroelectric plant, controlled by Eletronorte, a subsidiary of Eletrobras. CVRD is also the major shareholder in two other projects, Albras and Alunorte (developed in joint-venture with Nippon Al. Ltd.), to produce alumina and aluminium, and additionally it holds 46 per cent of another, the Trombetas Project which produces bauxite (Votorantim, a national group, holds 10 per cent, Alcan 24 and Billington 10). So far, in part due to the world crisis, there is only one project exclusively in the hands of a foreign corporation - Alumar (Alcoa and Billington).

As a consequence, so far CVRD is the one enterprise that benefits most from the state's strategy for mining development, a strategy that involves: (a) the definition of a 900,000 km^2 area for incentives (income-tax exemption for ten years, exemption of taxes upon equipment imports and upon sales); (b) infrastructural logistics (combined river and railway transportation, urban centres); (c) production of the basic support for aluminium production - subsidized electric power - a gigantic scale, which is consistent with the magnitude of resources involved, by the Tucurui Hydroelectric power plant.

In this process, the state's external debt increases, however, and it loses a substantial portion of its functions as a planner, inasmuch as part of the large projects are self-managed enclaves. In a sense there has been a return to a model similar to the one current in colonial times, based upon large self-supporting units of production and consumption. Labour market segmentation is evident. Company towns lodge the management board, the technicians (including those who warrant communication with foreign regions and countries), and are also a place for skilled manpower training, while in their surroundings grow satellite villages where unskilled mobile manpower have their dwellings.

In the process of global/fragmented space production, fragmentation widens with the private appropriation of large tracts of land by corporations that relate to a transnational spatial order of its own, dissociated from the national spatial order. This fact leads to the aggravation of contradictions between common and private interests, a statement that also applies to state-owned enterprises. Paradoxically, however, their expansion, provided by the state, is fundamental to nation-state building.

Localized Experiences: a Territorial Power

State strategies at national and transnational scale, although powerful, are not absolute and not so rational. The praxis of political struggle, represented by the multiplication of protest movements organized on a territorial basis that make claim for counter-spaces, brings into evidence the importance of the practical side in social space production, raising questions as to this theoretical construction.

The passivity of space users comes to an end. Although this theoretical construction, in its scope, acknowledges the conflicts and contradictions implicit in the social space production, the theoretical abstraction does not manage to handle the specific forms in which contradictions are manifest, nor the wide range of outcomes; that is, does not manage to handle the synthesis of new situations engendered. In other words, it does not account for the uncertainty which stems from specific aspects of the social body corresponding to processes current in other dimensions and scales, sometimes opposed to the dominating processes in a global scale, but that exert influence upon them.

It becomes necessary to analyse localized societies and social conflicts, raising the issue of regions as specific territorial powers, and of its possible influence on the direction taken by the social space production process. Will facts at a local scale - understood here not as micro but as localized - have the possibility of creating behaviour, and therefore capacity to generate an alternative political path to the established social space order (Becker, 1985c)?

Research experience in the frontier allows for the hypothesis that the resistance of small producers has an influential role in social space production and region formation. This role is played not only in armed conflicts, which break out from time to time in the struggle for land, but also in the conflicts that underlie everyday life, in the action and reaction chain that makes up the social relations and builds regions. Within these movements, the political potentiality of the frontier gains force.

It is thus possible to enrich the concept of region as a territorial power: its political force lies not only in hegemonic regional sectors, but also in non-hegemonic segments, whose role varies within localized social contexts.

Integration within a state-imposed spatial order is also integration within the global homogeneous space order but in which the logic of the whole is denied by the fragmentation of the details, as it is appropriated selectively in portions. Such appropriation in portions is nothing else than its territorial locations, which partially determine, in this way, the relative weights of class sectors or groups: the varying configuration of local societies - set of non-monopolist sectors - which make up the social expression of forming regions.

As a territorially organized society and specific historical reality, the region is a political power and may be understood as 'a territorial field that manifests the rise of a specific local power which the institutional frame simply legalizes' (Loinger, 1982). This assertion corresponds partially with the formation of new regions in the frontier, which result from the interaction of two territorial nets, at different levels: (1) the socio-political net, made up of the experienced space of social groups that actually are settled at the frontier: depending on its political potential - as a pressure group or, opposedly, as an object of manipulation - the experienced net is institutionalized, becoming the origin of new units of the federation, districts, municipalities and also states; (2) the technical-institutional net, made up of the territories appropriated and managed directly by the state in areas where there are strategic resources and/or real or potential conflicts. Some of these divisions are little more than a plan, but which nevertheless does not diminish their power of influence: they produce a new territorial division, simultaneous and conflicting with the official one, while not possessing the political and institutional means to guarantee people's representation. The previously existing space is thus reshaped in identifiable homogeneous/fragmented sets in various scales (Becker, 1985a).

The first homogeneous/fragmented set is identified at a regional-national scale, represented by the creation of legal Amazonia. The regional homogeneity is fragmented in two subregional homogeneous sets: Eastern Amazonia and Southern Amazonia. These subregional units, in turn, are fragmented in different local societies with uneven political potentiality, as shown in the cases of Eastern Amazonia and Rondonia.

1) Eastern Amazonia

Strategically located in the confluence of the dynamic centres in the central-southern region and the manpower reserves in Northeast Brazil, Eastern Amazonia was the first area to be occupied in the recent frontier expansion, having, as an axis for penetration, the Belem-Brasilia road. There, the state, through credit and subsidies, has clearly favoured accumulation by private farmers and economic groups. In order to achieve this goal, intense labour mobility has been stimulated to form a manpower reserve not only for local activities, but for expansion fronts as well.

Medium-scale landowners, cattle raisers coming from the southern-central region, make up the new dominant group, replacing the old hegemony of Brazil nut exploiters, and either disputing the local state apparatus with the traditional politicians or setting up alliances with them. A regionalist movement toward the replacement of the old local hegemony is supported by the action of large farmers and economic groups which, although not living in the area, exert pressure upon the central government to sever the region from its roots, through the creation of a new state, institutionalized as a federation unit.

Although there are a number of differentiated conflicts in the area, involving all scales and all actors, labour mobility stands as the main cause of conflict and resistance movements. In the continuous expropriation process, small producers achieve political consciousness, creating forms of resistance that include defence and attack tactics: invasion - of private and state land - and armed struggle. These movements trigger underlying conflicts between segments of the civil society, between this society and the politicians, between the Catholic Church and the state, and between small producers and wage-earners.

Although the resistance of small producers and rural workers in Eastern Amazonia has certainly influenced public opinion in favour of political freedom and land tenancy reform, local social structure conditions materialize in violent repression ordered by farmers and administrators of corporations, resulting in a large number of deaths.

(2) Rondonia (Southern Amazonia)

Located at the border with Bolivia, the present state of
Rondonia, for internal and foreign geopolitical reasons,[5]
has been the stage for a settlement strategy related to
state legitimation - the controlled distribution of land for
small producers, based in family work - involving one of the
most significant processes of social, political and economi-
cal change in the frontier, and generally in contemporary
Brazil.

Set off at the beginning of the 'seventies, this
settlement process was identified with the production of a
new region that was institutionalized as a state of the
federation. This production was guided and performed by the
federal government, but came into existence through the
hands of thousands of migrants, whose initiative, differ-
ently from Eastern Amazonia, has since become dominant.

The assertion of government hegemony was grounded on
the national security ideology, and found a basis in the
free land appropriation strategy. Its controlled distribution in official colonization projects has had an effect. A
small number of settlers was initially established with full
state assistance, in 100ha lots; a larger population
followed that 'spontaneously' established itself - the state
later legalized their situation.

The administration of the territory was directly
linked to the central government, through an appointed
governor and INCRA (National Institute for Land Tenancy
Reform and Colonization). The new local society and the
formation of dominant groups, in opposition to the traditional politicians linked to the rubber extractive economy,
was made through the political alignment of small producers:
some pioneers became connection elements between the local
society and the governor, directly or through the politicians, in exchange for personal favours and for the institutionalization of towns, districts and municipalities.

The intensive demographic growth (from 11,064 to
888,430 inhabitants between 1970 and 1984) and the 1982
elections made the transformation of the territory into a
state (in 1981) politically interesting. The systematic
build-up of the local state apparatus, closely related to
urbanization, has fostered change: graduate professionals,
from the south, hired to take charge of the new secretariats
at state and municipality levels, brought about the formation of a techno-bureaucracy that acquired land, associated
with large producers, and ultimately displaced the pioneers
from local power.

The resistance of small producers has played a primary
role in this process, appearing in two forms. (a) Systematic social pressure, represented by the spontaneous/induced
settlement in the context of state strategy and control that
forced the state quickly to alter the settlement forms

involved; the labour market organization with the family and
the share-croppers (a system that offered the possibility of
absorbing much larger numbers of producers in each demarcated lot); the multiplication and growth of towns built
exclusively by migrants, who so far remain as squatters; the
organization of squatters and small producers in associations, with support of the Church and political parties,
that seek to get rid of tradesmen's control. (b) In conflicts generated by unforeseen and non-state controlled
spontaneous/induced settlement. Recently population inflow
has far exceeded the state's control capacity, which is
reflected in conflicts and land invasion. Favoured by
population density, the population carries out organized
invasions, with tactics based upon information on the actual
legal situation of land appropriated by farmers and ven-
tures, invading areas with doubtful property title situation, and so having, in general, a real chance of achieving
their intents in the very field of official jurisdiction.

In Rondonia, popular resistance is not only characterized by the predominance of armed struggle, but also by a
systematic action toward the conquest of their own space,
and, nowadays, for its preservation. The influence this has
been having upon the state and the formation of the region
is noticeable, to the point of reaching an intense mobilization of civil society and the local state apparatus in
favour of land tenancy reform.

It is thus necessary to recognize a rising local power
of small producers whose struggle is to conquer and preserve
spaces according to the established rules - they want to
dispose of the land, to cultivate it, sell it, rent it or
even (as it has been registered in the area) divide it into
lots, aiming to create villages. In their ways of acting,
they show a great ability to adopt and improve the official
models, as well as to innovate and organize for the defense
of their rights, being therefore able to influence the
direction taken by the social space production process.

Final Remarks

Nation-state is a concept that refers to a continuous
historical process and not to a ready-made form of societal
organization. There is an explicit reciprocity between
nation and state which, however, does not necessarily entail
symmetric development: there may be larger or lesser
discontinuities between the consolidation of authority and
the incorporation of the state's subjects in the political
arena. There are, therefore, different ideologies of
nation-state - liberal-bourgeois, authoritarian - with
variable political routes to modernity. In Brazil, the
historical path was the authoritarian promotion of modernization, under the aegis of the state. State-building went

far ahead of nation-building, for no corresponding widening of the political arena in order to integrate all social segments took place. Nevertheless, modernization brought about new social actors that support state legitimacy and today show signs of getting more autonomy.

Spatial strategy and regional development policy were basic tools for nation-state building in the last twenty years. State-building experienced an impressive expansion through: the extension of the public sector into the hinterlands; growth and diversification of state enterprises; extension of institutional networks and fiscal control; control over land and over territorial redistribution of investments and population. As to nation-building, improved communication systems brought to large segments of the population a sense of membership in a broader territorial whole; state action entailed a minimum provision of social rights to rural populations (including private education, health services and retirement benefits) and by these means, through tutored incorporation, converted citizenship rights into authority's favours. Today, social movements against land expropriation are movements towards integration into the nation.

The fact that state-building went ahead of nation-building entails pragmatic consequences: stateness is far advanced not only as an idea but also as an institution, and it will be difficult to deal with the social and economic complexity of contemporary Brazilian society without state intervention. Solutions of social issues ought to be carried on at the national scale under governmental management and therefore they do not entail a reduction of state intervention but involve a change in the nature of this intervention; due to the close association between state and private capital, any change of the internal accumulation pattern must be accomplished through the state.

The new trends of national territory planning reflect the new influences in state action, which strive to fit together a national development policy and a response to the various social and political demands. Two trends are outlined at a national level - continued modernization, along with the atomization of state action, meeting municipal and regional claims.

For Amazonia, a global strategy called the Amazonia project has been under consideration, through the evaluation and integration of previous programmes; simultaneously, an economic ecological zoning has been suggested, together with closer attention to the economic usage of floodplains and the onset of a steel industry in the region (SIDERAMA), linked to SIDERBRAS, a holding of state companies for the steel sector. However, previous programmes have not been cut off, so that there is risk of engendering new irrationalities in territorial organization.

On the other hand, Amazonia is not the same any more. There is a new regional reality, bringing new problems, particularly regarding political representation. Considering that an authority appropriation has been carried out, it is our opinion that it is necessary to strengthen states and municipalities, expressions of social forces that have withered with centralization. This reinforcement does not seem to be enough to ensure a more democratic rule though: there are new actors on stage, and they must be acknowledged. Large corporations - private and state-owned - must be integrated to the region, having their rights and, above all, duties clearly identified; on the other hand, it is necessary to think of other ways to warrant political representation for the new local societies under formation. Territorial redivision can also be foreseen, with eventual creation of new official unities, not new states that would benefit large farmers but territories under central government jurisdiction, due to the need for a more solid political representation, for a more rational management, and for enforcing the law so as to curb arbitrary action by private concerns. Finally, it is fair to raise doubts as to the validity of keeping the large institution for regional development (SUDAM) at work, as it is an agency that manages an ideological region belonging to the past, which no longer exists.

Notes

1. Field research for part of a project financed by FINEP, CNPQ and CEPG, was coordinated by the author and carried out with the participation of Lia O. Machado and Mariana P. Miranda, and also with students from the Geography Department of UFRJ.

2. A more detailed analysis concerning these conditions may be found in Becker, 1985a.

3. The frontier occupation strategy in a context of urbanization is explicit in the papers of the consulting firm hired by SUDAM: 'Instead of the classical development method for unpeopled areas, based upon the prospecting of natural resources and subsequent settlement of the population to exploit them, it was proposed, conversely, first to urbanize the region, so that, once the people were settled, they would do the prospecting and necessary adaptations to exploit the resources themselves' (Racionero, 1978).

4. Transnational or multinational corporations are enterprises with at least two affiliates and ten per cent of their activities outside the country.

5. More detailed analysis of the importance of geopolitical factors in Amazonia and Rondonia are part of Becker 1982 and 1985c, respectively.

References

Becker, B.K. (1982) Geopolítica da Amazonia, Zahar, Rio de Janeiro

Becker, B.K. (1985a) 'The Frontier at the End of the XX Century: Eight Propositions for a Debate on Brazilian Amazonia' in International Economic Restructuring and the Territorial Community, Unido, Vienna.

Becker, B.K. (1985b) 'The Crisis of the State and the Region: Regional Planning Questioned', Environment and Planning D: Society and Space, 3,

Becker, B.K. (1985c) 'Frontiera e Urbanizacao Repensadas', Revista Brasileira de Geografia, 47: no. 3-4.

Becker, B.K. (1985d) 'Spontaneous/Induced Rural Settlement in Brazil', Working Paper, Habitat, United Nations, Nairobi.

Castells, M. (1985) 'Technological Change, Economic Restructuring and the Special Division of Labor' in International Economic Restructuring and the Territorial Community, Unido, Vienna.

Gaudemar, J.P. (1976) Mobilité du Travail et Accumulation du Capital, Maspero, Paris.

Lefebvre, H. (1978) De l'Etat, Union Générale, Paris.

Loinger, G. (1983) 'Une Economie Politique du Spatial et du Territorial', Espaces et Sociétés, 43.

Racionero, L. (1978) Sistemas de Ciudades y Ordenación del Territorio, Alianza Universidad, Barcelona.

Velho, O. (1985) 'Seven Equivocal Thesis on Brazilian Amazonia', Environment and Planning D: Society and Space, 3: 2.

Chapter 4

MARXISM AND SELF-DETERMINATION: THE CASE OF BURGENLAND, 1919

Andrew Burghardt

Marxism has been a close partner of self-determination throughout this century, and particularly since the end of World War II. The connection between them has, of course, been noted by many commentators. It is no secret that it is precisely this connection which has made some Western governments very hesitant about supporting movements for self-determination.

I will not analyse the reasons for this linkage. It is clear, however, that Marxist explanations prove to be highly attractive to groups which feel themselves to be downtrodden. Organisationally, the hierarchical structure which always seems to result, often with a charismatic leader in control, lends itself to the development of a dedicated, disciplined, and effective political force.

Dov Ronan has attempted to clarify the study of self-determination by postulating the existence of five archetypal forms: 1) nineteenth century, e.g. German and Italian; 2) Marxist class struggle; 3) minorities' self-determination associated with the ideas of Woodrow Wilson; 4) anti-colonialism; and 5) today's 'ethnic' quest for self-determination (Ronan, 1979). Whether or not such clear-cut distinctions can be made in the analysis of actual events is, of course, doubtful. Certainly Marxist class struggle, minority rights, and anti-colonialism seem to have become all mixed together in many of the recent movements for self-determination.

Ronan also sets up an 'us' and 'them' dichotomy, which he uses as the basis for discussion in a number of model cases illustrating his five types. In his analysis, this approach proves fruitful, and it will be made use of in this chapter. It suggests a simplified view of the world as being polarised, with an assumption of unity among both protagonists, 'us', and opponents, 'them'.

Burgenland: from Hungary to Austria

It is always useful to go back to the beginnings in order to come to understand how things came to be. In the case of Marxism and self-determination, it should prove enlightening to go back well beyond the contemporary post-colonial era, to the time after the First World War, when the boundaries between imperialism and colonialism lay in Central Europe, and not in Africa. The Burgenland case is especially suitable for such a study because it occurred at the high tide of Wilsonian self-determination, at a time, in fact, when self-determination was the officially proclaimed ideology of the world's rulers (Hula, 1984). The transfer of Burgenland from Hungary to Austria was one of the few territorial changes in which the popular ideal was allowed full play, unencumbered by major considerations of strategy or recompense. Equally important here, the decisions were made at a time when both Socialist and Communist governments were prominent in the area.

In November 1918, at the close of the disastrous war, the portion of West Hungary which became Burgenland was, economically and socially, in a quasi-feudal situation. Although most of the peasants were small landholders, the land use was overshadowed by the huge estates of the nobility, the Eszterházys, Batthyánys, Zichys, etc. Because of intensifying land-hunger and little local urban development, emigration was heavy. The only small city, Sopron (Odenburg), was essentially a marketing and administrative centre with little manufacturing. The large land holdings enclosed clusters of barracks, pseudo villages built by the landholders to house the contracted agricultural labourers. In the flat northeast of present Burgenland these encampments, or **pusztas** as they were commonly called, could house several hundred inhabitants. In later years these landless workers became supporters of the Communist Party.

The northern portions of this frontier strip were within fifty to one hundred kilometres of Vienna, and weekly commuting to jobs in Vienna had become common for the adult men by 1900. Vienna, as capital of the Austro-Hungarian Empire, was a rapidly growing city, with many job opportunities in unskilled positions. Northern Burgenlanders were soon predominant in the construction industry. During the work week they lived, several together in one room, in small apartments in the poorer districts of Vienna, while their families remained at home in the Burgenland villages. Also, a few textile industries located just inside the Hungarian boundary to take advantage of a cheap labour supply close to Vienna. Using Hroch's classification (see Chapter 1), northern Burgenland may be designated as one of those 'intermediate areas' where nationalism could develop, in contrast to the more traditional, peasant South, where it was unlikely to do so.

Political Parties in the Aftermath of War

In November 1918 both Austria and Hungary, which had just separated, had Socialist governments. In each case the Party controlled the capital city, but little of the rest of the country. In the election of February 1919, the Social Democratic Party of Austria obtained 70 per cent of the vote in Vienna (Pick, 1975), but only some 25 per cent in the rest of Austria - and most of that came from the iron and steel valleys of Styria and the city of Linz. Neither party had much practical knowledge of or contact with the peasantry. The Austrian party was headed by a galaxy of internationally-known figures, such as Karl Renner, Otto Bauer, Rudolf Hilferding, and Viktor, Max, and Friedrich Adler. The influential editor of the party newspaper, Die Arbeiterzeitung (The Workers' Paper), was a Burgenlander, Friedrich Austerlitz. (His American-born nephew, named after him, became famous in his own right as Fred Astaire.) Another Burgenlander, Julius Deutsch, was the founder of the Schutzbund, the party militia of the Social Democratic Party.

Ideological commitment to Socialism, both Marxist and Lassallean, was strong. Party discipline was stressed and usually maintained. Gaining control over the means of production was a central tenet, but a more important emotional factor in Central Europe was the anti-clericalism of the Party. By mutual agreement, it was assumed that one could not be a true Socialist and a devout Catholic at the same time. (When a founder of the Social Democratic Party, Viktor Adler, decided to become a Christian, he converted to Protestantism; Jacobs, 1985.) Since the Church no longer had large estates in Austria, church services were not interfered with, and religious education remained the norm, the conflict between the Party and the Church was perhaps more perceived than real. But the perception was certainly a powerful one.

For some of its dedicated members, socialism possessed religious qualities. Friedrich Adler stated 'I ... will confess that for myself socialism was a religious experience for me long before I knew of or understood its scientific doctrines'. The international aspects of socialism were compared to those of the Catholic Church, and 'the party program was viewed as the Eucharist, the manifest presence of truth and the means whereby the individual Social Democrat could partake of the spirit and matter of this truth' (Blum, 1985, pp. 149, 152). Karl Renner viewed the Catholic church as 'the first organising source of the public world' (Blum, 1985, p. 36).

The sense of a brotherhood of the working class was very strong. As a rule, as soon as a man took an industrial, transportation, or construction job, he automatically became a Socialist, and not only he but his whole family as well. They subscribed to **Die Arbeiterzeitung**, carried the red flag, and marched in the May Day parade - which is still a major feature of Viennese urban life. Consequently, most of Northern Burgenland developed into a Socialist stronghold, and has remained such ever since. The recent Austrian Chancellor, Fred Sinowatz, is from this area.

The political life of Austria thus became split between the two doctrinaire parties, the urban-centred Social Democrats, and the rural-based Christian (Sully, 1985). A brief attempt was made to have the two work together in a coalition government, but they soon split apart. Under the major economic and social pressures faced by Austria at the time, the two parties set up their own armed forces, leading to the brief, tragic civil war of February 1934 (Rabinbach, 1983).

The Bolshevik Revolution of 1917 made its impact felt as soon as World War One was over. It held a great attraction to the more doctrinaire members of the Austrian party, especially in April 1919 when, for a brief time, both Hungary and Bavaria were Soviet states. In an attempt to prevent the splitting of their party, the Austrian Socialist leaders tried to absorb the Leninist elements.

Otto Bauer distrusted the new Soviet Union, and felt that the dictatorship of the proletariat would transform itself into a dictatorship over the proletariat (Ziesel, 1985; Bourdet, 1968, Blum, 1985). Bauer had been a prisoner of war in Russia for three years. Perhaps because of his experience there, he favoured a 'slow' revolution, and 'defensive violence' only (Bottomore, 1978; Bourdet, 1968). He argued that 'like fascism [the Soviet Union] is the dictatorship of a ruling caste standing above the classes' (Rabinbach, 1983, p. 43). On theoretical grounds he felt 'the pretention of the Bolsheviks to install a dictatorship of the proletariat without a proletariat ... to be a dangerous illusion' (Bourdet, 1968, p. 65).

The Austro-Marxists did not want to surrender their democratic principles. This meant that the Austrian Communist Party was unable to take over the Social Democratic Party and hence remained insignificant (Klemperer, 1985), despite an attempted coup d'état on 15 June 1919, which cost twelve lives (Rabinbach, 1983). But it also meant that prominent members of the Social Democratic Party voiced much of the radical rhetoric, which was guaranteed to alienate large segments of the population.

In Hungary, on the other hand, the arrival of Bela Kun from Russia in November 1918 led to the Communist takeover of the Socialist Party, and the setting up of the Hungarian

Soviet in March 1919. In a time of chaos, when Hungary was being attacked by the Czechs, Serbs, and Romanians, the Communists tried to put their ideology into practice. There was a flight of refugees, churches were desecrated, and goods were expropriated from the countryside for the sake of the hungry proletariat of Budapest (Tokes, 1967). In the perception of the Western leaders, who were just then meeting in Paris to dictate the peace terms, Hungary had become the first expansion of the detested and feared Bolshevism out of Russia.

In both Austria and Hungary, then, internal political allegiances and conflicts became of far greater importance than questions of language or of ethnic adherence.

At this time there was, in fact, a group of men dedicated to the transfer of Burgenland to Austria on the basis of Wilsonian self-determination. They constituted a collection of individual efforts rather than a political party, at least until after the issue had been decided. They, and the Grossdeutsche Partei (Great German Party), which they subsequently introduced into Burgenland, were both non-Marxist and anti-Catholic. They envisaged the unity of all German-speaking peoples and, as such, saw the transfer of Burgenland to Austria as but one step in the achievement of an ultimate unity of all German lands. Their oft-repeated slogan and platform was 'Burgenland to German-Austria, and together with German-Austria to Great(er) Germany'. Perhaps inevitably, their emphasis on ethnic nationalism led them into Nazism. Because Burgenland did indeed come to Austria, and because they published extensively, this small group has been given more publicity by historians than they deserve. They had little to do with the important decisions, and were largely scorned or ignored. One Socialist leader referred to them as 'a bunch of beer hall table thumpers'.

In contrast, the Socialist leaders, who had the unwavering loyalty of almost half of the voters of Burgenland, expressed little sympathy for linguistic or cultural self-determination. For them, the attempt to unite peoples on the basis of ethnicity or language was a retrograde step. Max Adler stated caustically that, ' ... it has become clear ... that the idea of the nation is only an imperialist expression for the increased commercial and industrial power of a part of the nation, the ruling classes, for their supremacy in the world market, and thus for their world profits' (Bottomore, 1978, p. 127). As Karl Stadler has commented, 'German and West European Socialism had on the whole tended to ignore or underestimate the force of nationalism, regarding national movements as merely a by-product of the victory of capitalism over feudalism, and therefore essentially reactionary' (Stadler, 1971, p. 72). The many Jewish leaders of the Social Democratic Party were

strongly anti-Zionist (Jacobs, 1985). In their eyes the
bonding element was not to be language per se, nor religion
for that matter, but economic class. The fellowship was to
be one of workers united, whether they spoke German,
Hungarian, Czech, or Polish.

The leading Austro-Marxists, based as they were in
Vienna, favoured the supremacy of the German language. Their
discussions and publications were in German, their organisational structure functioned in German. Otto Bauer
thought ethnicity to be merely a temporary condition. He
drew a distinction between the Germans who were 'culture-
makers', and the Slavic nationalities within Austria, who
represented 'outdated heritages'. These latter groups
should assimilate themselves into the German nationality.
Despite being a Jew himself, Bauer refused requests of Jews
to have their own schools, because he maintained that their
only culture was that of the ghetto (Blum, 1985).

The Social Democratic Party position

The Social Democratic Party could not ignore for long the
nationalities question. The rising nationalist movements
posed a danger to the party in several ways: they divided
the members of the working class; they attracted the more
articulate members of the peasantry among whom the Social
Democratic Party hoped to gain support in its quest for
political power; and they held a strong attraction for
intellectuals and university students (Bottomore, 1978). The
first official statement on the issue was that made by the
Social Democratic Congress in Brunn (Brno) in 1899, which
tried to transcend the explosive territorial dimensions of
the nationalist strivings. Austria should be transformed
into a democratic federation of nationalities, but with the
emphasis on the individual.

The most influential position on the nationalities
question came to be that of Karl Renner. In his 1902 book,
Der Kampf der österreichischen Nationen um den Staat (The
struggle of the Austrian nations concerning the state),
Renner called for personal autonomy instead of territorial
autonomy, within the Empire. He referred back to Carolin-
gian precedents to support his 'personality principle'.
Nationality was not to be based on territorial units, which
he felt were the results of historic accidents, but rather
on membership in a corporation of individuals sharing the
same language and customs. Nationality would thus be
supranational (Blum, 1985). The Empire should be seen as a
Nationalitätenstaat (a state of nationalities), and could as
such serve as a model for the socialist organisation of the
whole world. The multi-national character of Austria was
thus valuable, in itself. There was therefore an immediate
need to establish those legal conditions which would achieve

these ends, and in so doing put an end to the political struggles for power among the nationalist movements (Bottomore, 1978).

In his statements of 1907 and 1908 Otto Bauer showed less sympathy for national differences than Renner had. For Bauer the nationality questions were clearly subordinate to those concerning the working classes. ' ... it was the duty of the working class to conduct the class struggle within the multi-national Empire' (Stadler, 1971), a Habsburg version of 'workers of the world unite', whatever language you speak.

It is obvious that Renner, Bauer, and the other leaders of Austro-Marxism were committed to the continued existence of the Austrian state, which then extended to Trieste and Lvov. As such they were against the territorial aspirations of nationalist leaders, not only because they divided the members of the working class, but also because they threatened to destroy the extended arena within which the Social Democratic Party was at work. The economic and political unity of Austria had to be maintained and fostered in order to secure that capitalistic evolution which was felt to be the essential precondition for the introduction of Marxist socialism (Jászi, 1929).

This is not to say that the socialist workers of northern Burgenland were unconcerned with the matter of which country they should be a part of once the Empire had disintegrated. They were rather very much in favour of the transfer to Austria. But their reasons had little to do with ethnic considerations, although their use of German certainly made communication easier. The ideas of Woodrow Wilson were used as means rather than as ends. Their jobs were in Austria; their primary fellowship was with Austrian workers. They were active parts of the Viennese socialist movement. To return to Ronan's dichotomy, they had a strong sense of 'us', but only a weak sense of 'them', at least in this context. They were part of a political fellowship based in Vienna, but they had no strongly felt need to leave Hungary.

Elsewhere in Burgenland, away from the commuter belt and the textile plants, the peasant life reigned supreme. Emigration had served to skim off the most restless persons, leaving behind those most committed to the maintenance of the family plot of land. The peasants remained within their established world-view of land, family and village, closely tied to the local church. The clergy was strongly anti-Marxist and most of the peasants shared the belief of their priests that the Socialist Party was Godless and menacing. The intemperate remarks of some of the more radical socialists reinforced the feeling of having one's life-style and spiritual values threatened.

The Hungarian Soviet

Between March and August 1919 Hungary was under the Communist government led by Béla Kun. 'Red Vienna' might be dangerous, but at least it was countered by the strength of the Christian Party in the rest of Austria. In Hungary, however, the Communists were running the entire country. It was this more than anything else which swung peasant opinion over towards Austria. As Andrew János (1971) has explained the situation,

> motivated by proprietary instincts, and bolstered by the continued influence of the Church and the traditional elites in the villages, peasant small holders and share croppers refused to cooperate with the new regime, were reluctant to deliver food for the city [Budapest], and in general sabotaged the economic measures of the Soviet [Hungarian] government. When the latter attempted to enforce its policies, scores of villages rose in arms against the proletarian dictatorship (p. 86)

Iván Völgyes (1971) has reported that the desperate government gave

> its permission [for some units] to use force and occasional terror to suppress counter-revolutionary activities in the country. Only a few peasants were hanged but, of even more consequence, was the indiscriminate collection of so-called war indemnity, and the requisitioning of materials by different bands causing terror and fear among the peasants. Whenever an area showed any sign of opposition, the villages in such an area were forced to give up their livestock, grain, and other goods on threats of death (pp. 85-86).

In June 1919 one of those peasant revolts broke out in seven villages to the east of Sopron. It was quickly suppressed and forced to pay the penalties described above.

With its priorities set on feeding the urban proletariat, and with little knowledge of the peasantry or contact with them, the Hungarian government succeeded only in solidifying rural opposition. There had always been a distrust of urban bureaucrats in any case. For the people on the land the desire to get out from under the Communists happened to lead in the direction of ethnic unity, and a union with the neighbouring Austrian peasants. In Ronan's terms, of 'us' and 'them' again, they were far more certain of what and whom they were against, than of what and whom they were for.

Still, despite the fact that most of the peasants of present Burgenland had German as their mother tongue, there would have been no transfer of territory if such had not been requested by the Austrian government. It fell therefore to the head of the Austrian delegation to the peace conference, Chancellor Karl Renner, to make the request. In his statement Renner pointed out the ethnic linkage of these border people with the Austrians, but he stressed in particular the need for Vienna and Graz to have an adequate food supply. Much like Béla Kun's concerns for the workers of Budapest, Renner faced the need to feed the workers of Vienna. However, remaining true to the democratic side of his Social Democratic Party, Renner asked not for an outright cession but rather that a plebiscite be held. The Western Allies, however, decided that a plebiscite was needless in view of the clearly defined ethnic frontier (Burghardt, 1962). It was thus the Allied decision makers more than the Austrians who relied upon the principle of ethnic self-determination to settle the matter. It was true too, though, that Hungary was Communist, and in a chaotic condition. The Entente made its decision on July 11, 1919, one month before the fall of Béla Kun and his replacement by the arch-conservative, Admiral Horthy.

The matter was not ended yet. The final peace treaties did not go into effect until two years later, and in that time the Hungarians had mounted a strenuous effort to keep control at least of Sopron, the only city in the ceded area. After some border clashes, the Allies arranged a plebiscite to be held in December 1921. In this plebiscite, the city voted to remain within Hungary. One of the strongest arguments affecting the vote was evidently the Marxism of the two governments. As one resident put it, the principal motive for the decision was a soberly realistic consideration. The conservative burghers of the city had no sympathy for the Austrian Marxism which at that time stood in the foreground (Burghardt, 1962). The argument made was that whereas Hungary had overcome Communism and was ruled by a 'Christian' government, the 'Red' Socialists were either in or very close to power in Austria.

Conclusions

Thus, paradoxically, Marxism played a prominent role in the transfer of Burgenland from Hungary to Austria, although the move is often seen as an expression of the 1919 obsession with ethnic self-determination felt by the Entente powers. I say paradoxically because it is clear that the ruling Marxists in both Hungary and Austria were not only not in favour of territorial ethnic self-determination, but rather disapproved of it as being retrograde and a hindrance to the coming unity of the proletariat of all ethnic groups.

Yet, ironically, it was the Marxist Socialists of Austria who made it possible for the aims of self-determination for the German-speaking Burgenlanders to be satisfied. They did this in three ways. First, by creating a strong sense of proletariat brotherhood, they provided a linkage between the northern Burgenlanders and Vienna, which was far stronger than any linkage created by a similarity of mother tongue. The great emotions were raised by class, not ethnic, considerations. Second, the Party supplied the organisational structure (Blum, 1985). Through the workers' councils, the hierarchy itself, and the influential newspapers, the Party made concerned action possible. Using Tonnies' terms, the <u>Gesellschaft</u> became the <u>Gemeinschaft</u>, the association became the community (Tonnies, 1957; Weber, 1978; see also Chapter 1). Whereas it was almost impossible to organise the smallholding landed peasants, the workers were organised from the beginning. And third, of course, it was the head of the Socialist Party who made the request to the Entente, and who set forth the arguments in favour of the cession.

In Burgenland the existence of a 'competing ideology' (referred to by Herman van der Wusten in Chapter 12), assisted rather than hindered the achievement of nationalist aims. 'Self-determination' may have formed the theoretical base for the transfer to Austria, but Austro-Marxism provided the structure and leadership which made its achievement possible. As has been proven to be the case elsewhere in more recent decades, the combination of Marxism and self-determination can create a very powerful political force.

The fact that Karl Renner made the request (and thus was taking land away from Marxist colleagues in Hungary), may be attributed largely to that primary concern of most Marxist governments, the need to feed the urban proletariat. In times of crisis, few matters can challenge the primacy of maintaining a secure food supply for the urban workers.

Thus, to return to Ronan's five archetypal forms of self-determination, Marxist class struggle, while seeming to work at cross purposes with it, was nevertheless interwoven with Wilsonian self-determination. Beyond that, one must say that Marxism had a very strongly polarising affect on the population. For its adherents it created a sense of 'us' far greater than any other unifying principle; for those opposed to it it created a sense of 'them' which was far greater than any sense of 'us'.

Postscript

As a postscript, there remains one more aspect to examine. Burgenland also contained 45,000 Croats, roughly 15 per cent of the population, speaking what has been termed by Colin Williams 'a subordinate language' (Chapter 13). Having been a minority within Hungary, they now became a minority within a consciously-German Austria. Frightened by the strident remarks of the pan-German nationalists, they began making their own cultural demands. Although broadly dispersed, slightly more than half the Croats lived in the commuter belt of the north, and the other half in the more typically peasant villages of the south.

The Croats of the north had become socialists alongside their German-speaking neighbours. Their reaction to this Croatian agitation was an outright denunciation of it. The Croatian intellectuals then decided to form their own political party, and contested the election of 1923. The Burgenland Croatian Party received twenty-five hundred votes in total, but only 272 came from the socialist north. The largest Croatian village in the province did not cast one vote for this party (Burghardt, 1962). The Croatian Party then opted to join the Christian Party, and the Croatian vote has been split on a north-south basis ever since.

No territorial changes could be envisaged, since these would have included the highly unpopular prospect of some form of union with Yugoslavia. Without a viable, compact territorial base, this weak movement for cultural self-determination has proven to be incapable of overcoming the strength of the Marxist 'competing ideology'.

References

Bauer, Otto (1907) Die Nationalitätenfrage und die Sozial demokratie (The Nationalities Question and Social Democracy), Volksbuchhandlung, Vienna.

Bauer, Otto (1908) Die nationalitätenfrage und der Staat The Nationalities Question and the State), Volksbuchhandlung, Vienna.

Bauer, Otto (1919) Der Weg zum Sozialismus (The Road to Socialism), Verlagsgenossenschaft 'Freiheit', Berlin.

Blum, Mark (1985) The Austro-Marxists 1890-1918, A Psychological Study, Univ. of Kentucky Press, Lexington, KY.

Bottomore, Thomas, and Goode, Patrick (1978) Austro-Marxism, Texts Translated and Edited, Clarendon Press, Oxford.

Bourdet, Yvon (1968) Otto Bauer et la Révolution, Textes choisis, présentés et annotés, Études et Documentation Internationales, Paris.

Burghardt, Andrew F (1962)   Borderland, A Historical and Geographical Study of Burgenland, Austria, Univ. of Wisconsin Press, Madison, WI.

Hula, Erich (1984)   Nationalism and Internationalism, European and American Perspectives, Univ. Press of America, Lanham, MD.

Jacobs, Jack (1985) 'Austrian Social Democracy and the Jewish Question in the First Republic', in Rabinbach, (ed.) The Austrian Socialist Experiment, Social Democracy and Austromarxism, 1918-1934, Westview Press, Boulder, CO.

János, Andrew C (1971)   'The Agrarian Opposition at the National Congress of Councils', in, János, Andrew and Slottman, William (eds.) Revolution in Perspective, Essays on the Hungarian Soviet Republic of 1919, Univ. of Calif. Press, Berkeley, CA.

Jászi, Oscar, (1929)   The Dissolution of the Habsburg Empire, Univ. of Chicago Press, Chicago, IL.

Klemperer, Klemens von (1985) 'The Habsburg Heritage: Some Pointers for the Study of the First Austrian Republic', in A. Rabinbach (ed.) The Austrian Socialist Experiment.

Paust, Jordan J (1980) 'Self Determination: A Definitional Focus', in, Alexander, Yonak, and Friedlander, Robert (eds.) Self Determination: National, Regional, and Global Dimensions, Westview Press, Boulder, CO.

Pick, Robert (1975)   The Last Days of Imperial Vienna, Weidenfeld and Nicolson, London.

Rabinbach, Anson ed. (1983) The Crisis of Austrian Social ism, from Red Vienna to Civil War, 1927-1934, Univ. of Chicago Press, Chicago and London.

Renner, Karl (1902) Der Kampf der österreichischen Nationen um der Staat (The Struggle of the Austrian Nations Concerning the State), Franz Deuticke, Leipzig.

Renner, Karl (1949)   The Institutions of Private Law, and their Social Functions, O. Kahn-Freund (ed.), Routledge and Kegan Paul Ltd., London.

Ronan, Dov (1979)   The Quest for Self-Determination, Yale Univ. Press, New Haven, CN.

Stadler, Karl R (1971) Austria, Praeger Publ., New York.

Sully, Melanie A (1985) 'Social Democracy and the Political Culture of the First Republic', in A. Rabinbach, The Crisis of Austrian Socialism.

Tokes, Rudolf L (1967) Béla Kun and the Hungarian Soviet Republic, Hoover Institute, Stanford, CA.

Tonnies, Ferdinand (1957) Community & Society, (Gemeinschaft und Gesellschaft), (Transl. & ed. by Charles P. Loomis), Michigan State University Press, E. Lansing, MI.

Volgyes, Ivan (1971) Hungary in Revolution, 1918-19, Univ. of Nebraska Press, Lincoln, NB.

Weber, Max (1978) Economy & Society, An Outline of Interpretive Sociology (ed. by Guenther Roth and Claus Wittich), Univ. of California Press, Berkeley, CA.

Ziesel, Hans (1985) 'The Austromarxists in 'Red' Vienna: Reflections and Recollections', in A. Rabinbach, The Crisis of Austrian Socialism.

Chapter 5

TOWARD A GEOGRAPHY OF PEACE IN AFRICA:
REDEFINING SUB-STATE SELF-DETERMINATION RIGHTS

Josiah A.M. Cobbah

> One does not point to one's hometown with one's left
> hand.
>
> An Akan Proverb

> (T)he common practice of opposing ethnicism (or
> pejoratively tribalism) ... overlooks the vital, if
> indirect, contribution that ethnic nationalism can,
> and has made to the growth and persistence of territo-
> rial state-wide nationalism.
>
> Anthony D. Smith

Introduction: Self-Determination in Africa, the Liberal Prescription

The decolonization period was no doubt the high point of the
principle of self-determination in international law. Within
a relatively short period after World War II, the vast
majority of European colonies in Africa attained their
independence. Having applied the principle of self-determination to European territories after World War I, the
Western powers found their hands being forced again, this
time by the pressure of the decolonization movement.

For the emerging African countries, however, the
opportunity for self-determination came with a condition.
Self-determination did not mean a dismantling of already
established colonial boundaries. Independence essentially
meant the replacement of a colonial administration with an
African one, and the duty of the outgoing colonial power was
simply to put into place constitutional traditions to create
Western-style liberal democracies. One British bureaucrat
had the following recommendation:

> (a)   system of official or bureaucratic government
> must inevitably be the first step in the development
> of communities such as those which the majority of
> colonies are composed. They lack the social organiza-
> tion which has been responsible for the evolution of
> representative institutions or forms of popular
> government in western countries ... They have not as
> yet those ties which make different elements in the
> community ready to subordinate sectional interest to a
> common purpose. (Hailey, 1944, p. 45)

Politicians and scholars felt that the ascriptive centrifugal forces of ethnicity could be overridden by the centripetal forces of the centralized state.   Thus the Mandate
System demanded a tutelage and gradual nurturing of the
colonial peoples' ability to govern themselves according to
the criteria of European liberal democracies: the achievement of democratic order, and the protection of minorities
(Ofuatey-Kodjoe, 1977, p. 87).

Two or three decades after the massive decolonization
programme, the liberal blueprint for self-determination is a
rarity in Africa.   In many parts of the continent, self-
determination has not guaranteed equality and peace.   The
continent is fraught with ethnic conflicts of all sorts, and
African leaders have been putting a lot of energy into
dealing with ethnic strife.   In some cases, where peaceful
measures have failed, politicians have sometimes violently
suppressed interethnic disagreements.   Ghana's Kwame Nkrumah
expressed the typical contempt for ethnic identification by
African leaders:

> (t)ribalism arose from colonialism which exploited
> feudal and tribal survivals to combat the growth of
> national liberation movements ... In the era of
> neocolonization, tribalism is exploited by the
> bourgeois ruling classes as an instrument of power
> politics, and as useful outlets for the discontent of
> the masses. (Nkrumah, 1970, p. 59)

Kwame Nkrumah and others who espoused socialism considered
ethnic identity as a bothersome obstruction that disturbs
the progress toward class-consciousness (Smith, 1983, p.
126).   For its advocates, the liberal state will eventually
render ancient customs and myths of common ancestry obsolete
and eventually citizenship of the state will overcome all
ethnic identities (Smith, 1983, p. 2).

Africans are still clinging to their ethnic identities, however, and if over twenty years of statehood
constitutes enough basis for drawing conclusions, then the
proponents of the unitary state may be overstating their
case.   Indeed, some scholars now contend that the tendency

of Western social scientists to interpret the African experience in terms of European history misses the important fact that the African world-view is one that is fundamentally different in kind from the European view - that is, whereas Europeans are conditioned to see themselves as independent individuals, Africans have always identified the 'self' in the context of their kinship ties (Akbar, 1984).

The expectation that African states will develop along Western lines may explain why the literature on statehood in Africa has so far been short on workable techniques for conflict reduction and resolution. What is available most times ignores the African reality. Most of the literature simply clings to a theory of modernization positing that as these countries become modernized (i.e. Westernized) ethnicity will die down and disappear. The resilience of ethnic identity in Africa and the problem ethnicity poses to class analysis are usually disregarded by analysts who are bent on replicating the Western experience in Africa.

In the blind acceptance of modernization theories, the more fashionable trend in contemporary geographical analysis is to elevate spatial function (interconnectiveness) over territory and in the process portray the state in Africa as a forum of free-floating individuals who seek to maximize utility within the unitary and progressively-integrating state. Fortunately, the geographic tradition has within it a 'sense of place' school of thought, one which we can fall on to satisfy the urgent need to apply techniques of conflict reduction and resolution to the continuing ethnic strife of Africa. This chapter seeks to make such a contribution.

First, the chapter seeks to explain the territorial nature of the African group identity. It then argues that this sense of ethnic identity is not necessarily negative and indeed shows a potential of becoming a force for development and self-government even within the present boundaries of African states. A call is then made for re-examining the meaning of self-determination in African states with a proposal for the territorial decentralization of these states.

The philosophy of Western liberalism undermines the vitality of human groupings that are intermediate between the state and the individual. The state has legal primacy in regulating the behaviour of its citizens and whatever authority groups have is held to be granted by the state (Frug, 1980). The legal primacy of the state is sanctioned under both domestic law and international law. Under this climate of assumed state sovereignty, the extreme political solution of ethnic secession is not a viable alternative for many groups. There is the need, therefore, for creative accommodative policies to deal with ethnic strife.

Africans and Geoethnicity

The African sense of being has been summed up by one philosopher to be: 'I am because we are, and because we are therefore I am' (Mbiti, 1974, p. 14). As mentioned earlier on, this sense of self can be distinguished from another which emphasizes individuals' obligations to themselves and the notions of natural rights and social contract which underlie the concept of citizenship (Cobbah, 1986; Nobles, 1976). African cosmology is built on the groupness of the community, sameness, and homogeneity. In fact, an increasing number of black psychologists are now calling for a reappraisal of the applicability of Western notions of individual actualization to African cultures (Nobles, 1976; Akbar, 1984, p. 394). For the African, group identity may implicitly be more important than individual identity (Foster, 1983).

Some writers have recognized the natural connection between extended family notions, kinship ties, and African ethnic identification. Horowitz, for example, recently remarked that '(t)he meaningfulness of ethnic identity derives from its birth connection - it came first - or from the acceptance by an ethnic group as if born into it. In this key respect (the primacy of birth), ethnicity and kinship are alike' (Horowitz, 1985, pp. 56-57). When one speaks of African ethnicity, therefore, one is speaking of birth relationships that form the very basis of the African's sense of being.

In Africa, this ethnic identity is above all other things a territorial identity. Nothing defines the ethnic group better than its 'standing place' (Knight, 1984). Thus the term geoethnicity has been used to describe the African ethnic phenomenon (Wai, 1983). Geoethnicity as opposed to non-territorial ethnic identification involves the historic identification of an ethnic group with a given territory, an attachment to a particular place, a sense of place as a symbol of being and identity. This geoethnic identity has been referred to as group politico-territorial identity (Knight, 1984).

African ethnicity is perhaps best expressed by the African attachment to the group's communal lands. The African chief and the elders in a community hold land in trust for the community. Meyerowitz had this to say about the Akan in Ghana:

> In practice the land in a state belonged, and still in the main belongs, to the various clans, whose chief gave the right of occupation to individual heads of families. The king, or Omanhene, was, and still is, the nominal holder of the land and might be called in to settle boundary disputes, but he had no right to

take or sell the land belonging to his subjects. If
he wished to grant land to refugees, he had to consult
the clan chiefs, for their agreement had to be secured
and it had to be decided on whose clan land the
refugees were to settle (Meyerowitz, 1958, p. 36).

Territory is therefore primary to African ethnic identity.
    The legitimacy of the state involves a perception of
state representation of ethnic or cultural identity, a
perception that Kelman calls an instrumental attachment to
the state (Kelman, 1969, pp. 176-288). For a stable
relationship to emerge between the state and the governed,
the governed need to feel some attachment and loyalty to the
state, accepting its authority as legitimate (Wai, 1983, p.
316). Where this attachment is absent or where a false
sense of unity is advanced, the state-governed relationship
becomes unstable. The evidence in Africa is that geoethnicity represents a system of primary sub-state attachments
which in some cases threaten the very idea of a unitary
state in Africa.
    Thus far, geoethnic identity in post-colonial Africa
has been seen in a negative light (Smith, 1983, p. 69). It
is my contention that geoethnicity, if properly understood
and nurtured, can become a force for socio-economic development in Africa. The territorial nature of ethnic identity
is a two-edged sword. Properly nurtured it can form the
basis of productive regional self-government and popular
participation in the affairs of the state. However, if the
present trend of pitching geoethnic aims against state aims
continues, ethnicity in Africa will be a source of strife
and even violence, without end.

The Positive Geoethnic Potential
Fishman captured the essence of African geoethnicity when he
remarked that kinship 'is the basis of one's felt bond to
one's own kind. It is the basis of one's right to presume
upon them in times of need. It is the basis of one's
dependency, sociability and intimacy with them as a matter
of course' (Fishman, 1976, p. 5). Research indicates that
even among the very poor urban inmigrants in contemporary
cities, ethnicity and kinship ties are relied upon to
provide much needed social security and assistance. Among
those who operate in the informal sector in Accra, Ghana, it
was found that dependency loads are very high, with kinsmen
usually being directly or indirectly responsible for the
well-being of newcomers into the city. Even the poorest
among these urban inmigrants were sending remittances back
to the hometown, were expected to retire back home, and saw
the stay in the city as an economic necessity. In effect,
they distinctively saw themselves in terms of their home-

lands (Cobbah, 1985). In the African city one finds various ethnic organizations, usually called improvement societies, and welfare unions dedicated to the financial and social benefit of ethnic group members in the city and to the improvement of the home territory (Little, 1965).

All these positive manifestations of ethnicity are usually overlooked, and ethnicity is viewed in a wholesale manner as a thorn in the flesh of the nation-state in Africa. Ethnicity should, however, be seen in the light of the fundamental communal nature of African existence: it is as rural as it is urban. Properly understood, geoethnicity reflects a sense of pride and identity in a given territory and examples abound in Africa of cases in which ethnic pride and the love of the homeland have translated into developmental efforts to uplift depressed regions.

There is no reason why African countries cannot utilize this ethnic pride in a positive manner This geoethnic force could be tapped to develop a decentralized system of governance through which citizens of African countries will actually participate in political decision-making. Once scholars and policy makers concede ethnicism, they can begin to fathom opportunities for African nationalism to operate simultaneously at different levels. Ethnic loyalty needs not always conflict with state loyalty. The two levels of loyalty need not conflict, and multiple loyalties are quite common in today's world; Coleman (1958) has applied the term concentric circles of allegiance to situations of multiple loyalties.

Human beings indeed have ties to different scales of territory and can operate at several levels of abstraction at any one time, from parochial regionalisms to broader nationalism (Knight, 1982, p. 515). If political systems react to these natural attachments realistically then citizens can have meaningful self-determination short of secession from the state.

Consociational Democracy: the African Possibilities

Models of consociational democracy seek to explain the paradox of coexistence of strong subcultural divisions with democratic stability. In his typology of political systems, Almond (1956) distinguished between the Anglo-American type of democracy and continental European political systems. Anglo-American systems are said to have a homogeneous, secular political culture and a highly differentiated role structure, 'in which governmental agencies, parties, interest groups, and the communication media have specialized functions and are autonomous, although interdependent' (Lijphart, 1969, p. 207). Anglo-American systems contrast with continental European democracies which are marked by 'a fragmentation of political culture with separate political sub-cultures' (Lijphart, 1969, p. 207). To avoid unintended geographical connotations, Lijphart classified these two systems as centripetal and centrifugal.

If this classification of political systems is applied to the African continent, we may reach the conclusion that African geoethnicity pulls us toward the centrifugal category. In spite of the geoethnic fragmentation in sub-Sarahan Africa, however, political institutions tend to be of the types that one would expect in centripetal systems. African countries continue to maintain highly centralized administrative patterns. As discussed earlier on, these patterns in most cases constitute a denial of the various cleavages that plague African societies and seek to affirm the dogma that as these countries modernize, ethnic cleavages will disappear.

To deal with the cleavages that characterize centripetal forces in the state-building, consociational democracies have been developed. Some important characteristics of consociational democracies are: 1) a commitment to accommodate divergent interests and demands of subcultures; 2) an ability to transcend cleavages through a common effort by rival subcultures; 3) a commitment to the maintenance of the system and to the improvement of its cohesion and stability; and 4) an understanding of the perils of political fragmentation (Lijphart, 1969, p. 216). A commitment to cooperation is essential. In the context of the African experience, consociational politics will mean that political arrangements should remain sensitive to geoethnic demands and endeavour to work these demands into the broader aspirations of the African state. This sensitivity is important in order to foster citizens' instrumental attachment to the state and to advance the legitimacy of the state. Moreover, where one accepts that there is a moral right of reasonable self-determination, such a sensitivity to geoethnic cleavages will mean that the state in Africa may see peace and stability in spite of its usually arbitrary and controversial political boundaries.

The centralization of politics in the African state has meant that many citizens, and especially rural citizens, are essentially shut out of political participation at the state level and even in matters that affect them only at the local level. There are good reasons for this centralization. In the drive to modernize and escape the Third World trap as fast as possible, the state had to seize control of the situation and utilize every available resource in order to achieve development ends. The decentralization systems that the colonial administrations left behind were found inadequate and primitive by the new nationalist governments and were eradicated. Thus in both former British and French colonies vestiges of indirect rule and native administrative patterns were done away with (Bakheit, 1971).

The desire to utilize the full range of state power for socio-economic development is also indicated by the increasing number of one-party states with national integration manifestos. In the quest to foster national unity, the political leadership usually preaches against or attempts to stamp out ethnic differences and ignores territorial cleavages. It is fair to say that so far African political development exhibits a tendency to avoid ethnic realities.

The experiences of Nigeria and the Sudan show us that conflict avoidance solutions and heavy-handed suppression of ethnic conflict usually fail to provide the desired political stability. It is my contention that it is only through a deliberate commitment to consociational politics and territorial self-determination respecting reasonable geoethnic aspirations that peace and stability can be achieved.

The Contents of an African Policy

It should be made clear that any realistic attempt to create a pluralistic democratic system which accepts and builds on the geoethnic nature of African societies will most likely run foul of the established tenets of Western liberalism. Liberalism upholds the rights of the individual against the state. Ethnic rights are however group rights which are not always reducible into individual rights. Whereas the state is functionally a collection of individual citizens, the ethnic group's shared identity supplies it with an ingredient whereby the group becomes mightier than the mere sum of the individuals.

I have argued elsewhere that any attempts to elevate the individual's importance over the group's identity in Africa will lead to a conflict situation because the seemingly 'detribalized' African does not hesitate to resort to his ethnic identification when faced with problems arising out of his relationship with the state (Cobbah, 1986). Moreover, the political reality of liberal politics in a unitary state has been that even where individual rights are constitutionally guaranteed, group affiliations (be they ethnic, racial or class associations) usually come into play, resulting in a need for affirmative action programmes (which protect groups) in order to guarantee minority citizens' individual rights. In effect states have had to resort to the sanctioning of 'positive unequal treatment' of minority groups in order to deal with the problem of 'negative unequal treatment'. Affirmative action programmes in the United States, although not territorial in nature, are an example of a deprived minority's 'positive unequal treatment' (Rhoodie, 1984, p. 73).

Successful consociational politics in Africa should be based on the geoethnic reality of the African state. A primary feature of African political practice should be a recognition of the ongoing tension between natural ethnic cleavages and the integrative tendencies of the contemporary African state. The modern state has been referred to as a territorially well-defined centralized unit (Tilly, 1975, p. 27). Likewise Smith accurately characterizes the likes of African states as entities whose 'governmental institutions claim total independence of other internal institutions and brook no rivals within [their] territorial domain' (Smith, 1983, p. 14). Thus assertive behaviour by sub-state entities such as the Ibos, the Eritreans, the Southern Sudanese non-Arabs and countless others arouses the displeasure of the state. The natural result of such displeasure has been either mass violence or state oppression of dissident groups. A reinterpretation of the right of self-determination is therefore necessary in Africa.

Sub-State Self-Determination

Neuberger has observed that:

> (i)n the twentieth century Third World, there is a crucial distinction between "colonial (or better, anticolonial) self-determination," which is the liberation of Asian or African peoples in a colony from European colonial rule, and secessionist self-determination, which represents a people's aspiration to break out of the post-colonial state and achieve liberation for one Afro-Asian people from rule by another Afro-Asian people (Neuberger, 1986, p. 7).

He further refers to anticolonial self-determination as 'grand self-determination' and to self-determination concerned with the internal politics of the state as 'small self-determination' (Neuberger, 1986, p. 8). In grand self-determination the 'self' refers to the 'colonial self'. Small self-determination in Africa however involves a challenge to the hegemony of the state and is therefore found to be threatening. This internal self-determination may come in the form of a demand for secession as in the Biafran case, but, as mentioned earlier, it is fair to say that in most African states today the state entity is a fait accompli and is accepted by the majority of sub-state ethnic groups. The individual's primary attachment is still to the ethnic groups, but it is probably an exaggeration to assert that nothing short of secession will satisfy ethnic demands of self-determination. A commitment by all factions to consociational democracy and to negotiated constitutional instrumentalities within the already defined multi-ethnic state offers a chance for peace and stability within a framework of ethnic self-determination.

Rhoodie (1984) has coined the term 'national condominium' to refer to a hypothetical country

> ... composed (constructed) of various blocs or units, i.e. races, ethnic communities or other groups. Each unit has a moral and legal right to continued existence. Each unit 'owns' such rights as the use of their own language, educational and legal system, religion, local and provincial government, etc. (Rhoodie, 1984, pp. 297-298).

Although the example of the South African Bantustan system indicates a strong potential for the abuse of a 'condominium' idea in a ranked society, it is contended that such an arrangement, where it is negotiated in the spirit of consociational fairness and equality, has a strong chance of success in Africa.

Within this consociational structure each geoethnic unit will enjoy rights which are internal to that particular unit. The state will consist then of independent geoethnic jurisdictions which are brought together only by specific concerns determined by consensus to fall under a broader state jurisdiction. Such a system calls for a rigid constitution which specifies inalienable geoethnic rights (within geoethnic territories) and which further distinguishes these ethnic specific rights from rights enjoyed in common by all citizens under state jurisdiction.

A written constitution cooperatively arrived at should spell out essential constitutional instrumentalities including: 1) a formal distribution of powers between geoethnic units and the state; 2) the establishment of an institution charged with the interpretation of the constitution; 3) a schema for legislative representation, 4) the requirements of geoethnic and state citizenship; and 5) a well-defined process for constitutional change (Livingston, 1956, pp. 9-11). These constitutional instrumentalities should reflect the nature of the geoethnic problem in a particular state.

In the typical African country the political system invariably has to deal with the realities of scarce economic resources and skilled manpower and the fact that there is usually a geographical imbalance in the distribution of such resources. The resource problem should be dealt with from a perspective which recognizes the fact that those on whose ancestral lands particular resources are found will naturally expect to derive more benefit from the resources than outsiders. If this reality is accepted then a balance could be struck between the use of resources for the benefit of the overall state and parochial geoethnic fiscal and economic development demands.

The problem of resources is a serious one. In the Sudan it has been argued that the embers of failure of the 1972 Addis Ababa agreement between the northern Arabs and the non-Muslim south were fanned to a large extent by the inability of the south to maintain itself economically and the refusal of the north to advance enough resources for the depressed south (Lesch, 1985, pp. 8-10). Ninsin (1980), a Ghanaian political scientist, is reported to have suggested that a reorganization of Ghana's political system should involve a democratic system which would require a combination of the socialization of the major means of production and the diffusion of economic and political power on the basis of smaller geopolitical units (reported in West Africa, Dec. 15, 1980, p. 2628). Ninsin's model seeks to avoid the emergence of a bureaucratic dictatorship by using 'a system of mass political and economic controls' (West Africa, Dec. 1980, p. 2628). It seems to me that the most effective means of achieving 'socialization of the means of production' is to depart from the present centralized, centre-down approaches and decentralize control over the fruits of territorial resources from the state capital to geoethnic centres.

Over the past two decades a number of African governments have announced a variety of decentralization programmes and a commitment to rural development. These programmes have on the whole had very little effect on the levels of rural depression on the continent (e.g. Cobbah, 1979). Much of the terminology used has been emotive in the sense that in most cases decentralization created toothless regional and local institutions which continued to look to the central bureaucratic elite of the state for regional sustenance. Because of the emotive labels usually associated with African decentralization 'one must question whose participation the architects of the reform had in mind' (Conyers, 1986, pp. 596). Usually these reforms are not meant to redistribute political or economic power in any real sense. Governments usually ignore their own plans or implement the plans in such a manner that the state's power is not diluted.

Taking Decentralization Seriously

The framework for territorial self-government within many African countries already exists in terms of geoethnic units. While it is true that these countries inherited international boundaries that disregarded the 'groupness' of many border populations, it is also true that within the various countries provincial administrative boundaries were as much as possible drawn along ethnic boundaries. This framework can form the basis of a sense of self-determination and self-expression within the present boundaries of African states.

In a lot of ways, the liberal state and especially its unitary nature is related to the notion that we can always make our subjective desires subservient to reason. The efficient conflict-free state is therefore seen as one in which individuals are rid of their subjective straitjackets and in which scientific imaginations of effort and competition thrive. In Africa, this has meant the suppression of ethnic identity and a reluctance to consider the feasibility of meaningful sub-state identities.

If we were to define democratic decentralization to mean popular involvement in the decision-making process, then geoethnicity can be seen as a step in the right direction. Frug has pointed out that, for genuine democracy to flourish, there must be a transfer of power to decentralized units: 'No one is likely to participate in the decision-making of an (territorial) entity of any size unless that participation will make a difference in his life' (Frug, 1980, p. 1070). Geoethnicity makes a difference in African lives and provides an avenue for popular participation.

Moreover, an emphasis on the territorial identity of citizens will have the effect of lessening tensions brought about by competition of ethnic groups at the state level. At present, matters which are entirely local in nature assume such national proportions because of the absence of local mechanisms and jurisdictions to address them. The decentralization required will involve a clear identification of issues that are country-wide in character and those that are local in nature. For example, in the Sudanese case, the 1972 agreement between the Arabized north and the African south left broad areas of socio-economic development and administration under local jurisdiction. This example of recognizing the potential for ethnic self-government and self-sufficiency within the accepted boundaries of the state has great potential and should be explored by African governments.

It is conceivable that decentralization in this geoethnic sense could mean that even in the area of foreign affairs the state will have to relinquish part of its traditionally unquestioned jurisdiction. For example, many African geoethnic units straddle state boundaries. In such cases the different states involved need to come to an accommodation with each other and involve particular ethnic groups in matters relating to free interstate movements and citizenship rights. In effect the geoethnic policy advocated here will have considerable international consequences. African countries will have to take another look at their usually rigid perceptions of their state boundaries.

Consociational politics recognize that majority rule (by the ethnic majority) may be oppressive and destabilizing and therefore calls for established formulae and agreements between sub-state groups. At the local level, there should be control over such sensitive areas as language and education and over local economic resources.

Development in African countries is still very lopsided and sometimes reflects the territorial priorities of the ruling elite. Since development is urban-based in the majority of these countries, the effect is that ethnic groups which are concentrated in the rural hinterlands remain relatively untouched by developmental efforts. The geoethnic policy being advocated here will mean that territorial groups will have a direct say in the use of local resources, and benefits that need to be shared will be considered within the spirit of consociational arrangements.

Africans take pride in their ethnic groups and have demonstrated the will and ability to channel this pride into the growth and survival of their homelands. What policy-makers should do is recognize this and reflect it. One can argue that in most African countries, the raison d'être exists now for people to see themselves, even if not yet totally, as Ghanaians, as Nigerians, or as Sierra Leoneans. The ethnic conflict that is being experienced in these countries calls for creative and responsible consociational measures.

Ranked versus Unranked Geoethnicity

In cases where geoethnicity has come to coincide with social class status, the extent of geoethnic self-determination may have to be greater and the commitment to consociationalism more total. In a ranked situation, a geoethnic group usually perceives a class cleavage which is coincidental with the geoethnic boundary. Unranked situations are those in which ethnic groups are parallel, with no group subordinate to another (Horowitz, 1985, p. 22).

Within the South African ranked situation one cannot by any means argue that the homelands fiasco comes anywhere close to the consociational democracy being advocated here. One ingredient of consociation is present, that is, South Africans of all races do see South Africa as their country. However, as Adam (1983) has pointed out, a realistic resolution of the political crisis is not possible as long as the white minority continues to impose a solution on the black majority. In this sense the homeland policy really constitutes a manipulation of ethnicity by a powerful minority to maintain the minority group's dominant position.

Where ethnic conflict has a class colouration, attempts at consociationalism will be more difficult and the territorial arrangements will have to be more far-reaching because of historical suspicions and animosities. Thus in the Sudan, the less developed southern provinces have harboured an historical resentment against the Arab northern region, and these resentments are being played out in the terms of what southern rebels will accept as adequate autonomy (Wai, 1983).

One point needs to be made. Ranked situations may be marked by protracted disagreements and even violence, and this may lead ruling elites to attempt the suppression of geoethnic tendencies instead of seeking a peaceful accommodation. Again a commitment to consociationalism is necessary on both sides.

Conclusion: A Geography of Peace in Africa

The post-colonial state in Africa has had as one of its major problems the task of keeping the state together. However, notwithstanding patterns of ethnic dissatisfaction with state policy and the many calls for secession, it is fair to say that modern international law is not about to react favourably to a broader interpretation of the concept of territorial self-determination. Through the near future, therefore, African states are going to remain with their present international boundaries. African countries therefore have to become creative.

In order to cut down on ethnic strife this chapter has called for a reinterpretation of the concept of self-determination in a manner that will grant to geoethnic groups considerable control over their local affairs and ensure a fair share of state resources. To achieve this self-determination, a policy of consociation is recommended, one which will respect the existence and equality of geoethnic groupings, and strive to turn the natural cleavages between ethnic groupings into sources of conflict resolution rather than perceiving these cleavages as a threat to the state.

The United Nations' Resolution 1514 (XV) of 1960 explicitly stated that:

> All peoples have the right to self-determination; by virtue of that right they may freely determine their political status and freely pursue their economic, social and cultural development

Self-determination in Africa has however been limited to the colonial state entities that African peoples inherited. Ethnicity has largely been frowned upon by scholars and the African political leadership. I started this chapter with a translation of an Akan proverb which says that a person does not point to his hometown with his left hand. The proverb means that anyone in his or her right mind will show attachment to the ethnic territory. Africans have shown that they have a tendency to point to their hometown only with the right hand. It is my view that the time has come to accept this fact for what is is and make it work for peace on the African continent. The ethnic 'self' remains the fundamental territorial 'self' in Africa. In effect, I do believe that geoethnicity should form the basis of an African geography of peace.

Acknowledgement

The financial support of the National Research Council in Washington, D.C., made it possible for the author to travel to the meeting of the Study Group on the World Political Map. The author is also grateful to Professor Bert B. Lockwood, Jr., the Director of the Urban Morgan Institute for Human Rights, University of Cincinnati College of Law, and to Mr. Mark Fitch for their intellectual support and encouragement.

References

Adam, Herbert (1983) 'The Manipulation of Ethnicity: South Africa in Comparative Perspective', in Donald Rothchild and Victor Olorunsola (eds.) State Versus Ethnic Claims: African Policy Dilemmas. Westview, Boulder, Colorado.

Akbar, N. (1984) 'Africentric Social Science for Human Liberation', Journal of Black Studies, 14, 395-414.

Almond, G.A. (1956) 'Comparative Political System', Journal of Politics, 18, 391-409.

Bakheit, Ga'Afar M.A. (1971) 'Native Administration in Africa', in Yusuf F. Hassan (ed.) Sudan in Africa. Khartoum University Press, Khartoum.

Cobbah, Josiah A.M. (1979) 'Development and Regional Inequalities in Ghana: 1960-1970', unpublished M.A. Thesis, Carleton University.

Cobbah, Josiah A.M. (1985) 'The Role of the Informal Sector in National Development and Integration Processes', unpublished Ph.D. Thesis, University of Cincinnati.

Cobbah, Josiah A.M. (1986) 'African Communalism, Liberal Ideology and Individual Rights: Where is Culture in the Human Rights Debate on Africa?', paper presented at the Seventh Annual International Human Rights Symposium and Research Conference, Center for the

Study of Human Rights, Columbia University, New York, New York, June 9-13, 1986.

Coleman, J.S. (1958) Nigeria: Background to Nationalism. University of California Press, Berkeley and Los Angeles.

Conyers, D. (1986) 'Future Directions of Development Studies: The Case of Decentralization', World Development, 14, 593-603.

Fishman, Joshua A. (1976) 'Language and Ethnicity', paper presented at Conference on Ethnicity in Eastern Europe, University of Washington.

Foster, Herbert J. (1983) 'African Patterns in the Afro-American Family', Journal of Black Studies, 14, 201-232.

Frug, Gerald E. (1980) 'The City as a Legal Concept' Harvard Law Review, 93, 1059-1154.

Hailey, Lord (1944) The Future of Colonial Peoples. Princeton University Press, Princeton.

Horowitz, D. (1985) Ethnic Groups in Conflict. University of California Press, Berkeley and Los Angeles.

Kelman, H. (1969) 'Patterns of Personal Involvement in the National System: A Social Psychological Analysis of Political Legitimacy', in J.M. Rosenau (ed.) International Politics and Foreign Policy. Free Press, New York, 276-288.

Knight, D.B. (1982) 'Identity and Territory: Geographical Perspectives on Nationalism and Regionalism', Annals of the Association of American Geographers, 71, 514-532.

Knight, D.B. (1984) 'Geographical Perspectives on Self-Determination', in P.J. Taylor and J.W. House (eds.) Political Geography: Recent Advances and Future Directions. Croom Helm, London, 168-190!

Lesch, A.M. (1985) 'Rebellion in the Sudan', UFSI Reports, no. 8.

Lijphart, A. (1969) 'Consociational Democracy', World Politics, 21, 207-225.

Little, K. (1965) West African Urbanization. Cambridge University Press, Cambridge.

Livingston, W.W. (1956) Federalism and Constitutional Change. Oxford University Press, London.

Mbiti, J.S. (1974) African Religions and Philosophy. Doubleday, New York.

Meyerowitz, E. (1958) The Akan of Ghana. Faber and Faber, London.

Neuberger, B. (1986) National Self-Determination in Post-Colonial Africa. Lynne Rienner Publishers, Boulder, Colorado.

Nkrumah, K. (1970) Class Struggle in Africa. International Publishers, New York.

Nobles, W.W. (1976) 'Extended Self: Rekindling the So-Called Negro Self-Concept', Journal of Black Psychology, 2, 15-24.

Ninsin, K. (1980) 'The New Democracy', (reported in West Africa 15 December 1980, p. 2628).

Ofuatey-Kodjoe, W. (1977) The Principle of Self-Determination in International Law. Nellen Publishing, New York.

Rhoodie, E. (1984) Discrimination in the Constitutions of the World. Brentwood, Columbus, Georgia.

Smith, A. (1983) State and Nation in The Third World. St. Martin's, New York.

Tilly, C. ed. (1975) The Formation of National States in Western Europe. Princeton University Press, Princeton.

United Nations (1960) Declaration on the Granting of Independence to Colonial Countries and Peoples, General Assembly Resolution 1514 (XV), 14 December 1960.

Wai, D. (1983) 'Geoethnicity and the Margin of Autonomy in the Sudan', in Donald Rothchild and Victor A. Olorunsola (eds.) State Versus Ethnic Claims: African Policy Dilemmas. Boulder, Colorado, 304-330.

Chapter 6

NATIONAL INTEGRATION PROBLEMS IN THE ARAB WORLD: THE CASE OF
SYRIA

Alasdair Drysdale

National integrative processes continue to be a central
concern of political geography (Williams, 1982; Kliot and
Waterman, 1983; Paddison, 1983; Clarke, Ley and Peach, 1984;
Taylor, 1985). However, most attention has focused on the
national unity problems of multi-ethnic societies in the
developed world, most of whose states have well-established
national identities. The Arab countries have attracted
little interest, perhaps because they are generally not
viewed as having severe national integration problems, with
the obvious exceptions of Lebanon, with its multifaceted
sectarian civil war, Sudan, with its racial strife, and
Iraq, with its Kurdish problem. Moreover, in comparison
with sub-Saharan Africa and other Third World areas, the
Arab world is relatively homogeneous culturally, with Arabs
and Muslims accounting for the vast majority of its inhabitants. In reality, many Arab states have integration
problems, and these are of interest because they raise
questions about the meaning of many of the words and
concepts political geographers routinely use to describe
integration processes. Such everyday terms as nation,
nationalism, national identity, separatism, partition, and
regionalism can, in the context of the Arab world, be
ambiguous and misleading. Rather than attempt to examine
the integration problems of all the Arab countries, an
impossible task in such a brief paper, specific reference
will be made here to Syria, whose particular state-building
experience illustrates in some form most of the problems
found within the region as a whole.

National Identity

Most states in the Middle East were created in the nineteenth or twentieth centuries by the colonial powers, who
ignored underlying cultural patterns when they drew the
political map (Drysdale and Blake, 1985, 41-74). The

division of the Arab world, much of which had enjoyed a certain unity under the Ottoman Empire, coincided with the rise of Arab nationalism, whose principal goal was the independence and unity of the Arab nation (Haim, 1964). The partition (taqsim) of the Arabs into a number of states was widely viewed by those who inherited the political map as evidence of imperialist divide et impera policies. Arab nationalists portrayed the boundaries that separated them as wholly artificial and arbitrary. In their eyes, the states themselves lacked legitimacy because of their origins and were destined to be replaced in the future with an Arab state that would extend from the Atlantic to the Indian Ocean. By definition, there could be no Egyptian or Syrian or Iraqi or any equivalent nationalism because these were not distinct nations, but components of one single Arab nation. As Lewis (1964, p. 94) noted, loyalty to the individual states, to the extent that it existed, was 'tacit' and 'surreptitious' and Arab unity was the 'sole publicly acceptable objective of statesmen and ideologues alike'. The term national identity was reserved for describing the Arabs as a whole. Arab nationalism is, therefore, an example of what Orridge (1982, p. 44) termed a 'unification' nationalism and similar in certain respects to nineteenth century German and Italian nationalism, which sought to unite the fragmented German and Italian culture regions into nation-states. To this day, Arabs think of themselves in greater or lesser degree as part of one nation (qawm) which is subdivided into many states (Rodinson, 1981). Although the dream of uniting the Arabs into one state is now recognized as unattainable, Arab nationalism still has considerable ideological potency.

In reality, the Arab world is diverse; its constituent parts have had unique historical experiences and differ politically, socially, economically, and even culturally to some degree. Consequently, Arab states have gradually acquired their own state-ideas and identities which are, in a sense, 'national'. In some instances, these derive in part from the states' historical continuity as distinct and relatively stable political geographic entities. Morocco's territorial roots, for example, go back several centuries, even if its current boundaries are recent. Egypt displays an even greater degree of spatial continuity and political geographic coherence, with a history that spans several millenia. Tunisia also boasts a long history as a separate entity, albeit often within the framework of a larger empire. Several countries are set apart by their cultural distinctiveness. Oman, for example, is unique in being largely Ibadi Muslim, while the Yemen Arab Republic differs in being predominantly Zaydi Muslim. The conservative Wahhabi Muslim movement has differentiated Saudi Arabia from its neighbours. Occasionally, 'national' self-awareness resulted from foreign oppression. Algeria's identity, for instance, was forged dialectically in opposition to the French, and Libya's in opposition to the Italians. In many cases, states have acquired a raison d'etre through simple inertia (Syria and Iraq) or through close association with a monarchy (many of the Gulf shaykhdoms). Arab nationalism has not achieved its most basic goal of uniting the Arabs, in large part because of growing attachments to these individual states (Ajami, 1978-79; Farah, 1986). This change was recognized by Iraq's President Saddam Husayn in a 1982 speech:

> Arab unity can only take place after a clear demarcation of borders between all countries ... Arab unity must not take place through the elimination of the ... national characteristics of any Arab country. The question of linking unity to the removal of boundaries is no longer acceptable ... Any Arab would have wished to see the Arab nation as one state ... But these are sheer dreams. The Arab reality is that the Arabs are not 22 states ... Unity must give strength to its partners, not cancel their national identity [emphasis added] (Foreign Broadcast Information Service (FBIS), 13 September 1982)

Because pan-Arabism has achieved so little, some scholars have portrayed it as moribund. For example, Ajami (1981, p. 127) notes that the Arab states 'are less shy about asserting their rights ... The Arabs who once seemed whole, both to themselves and to others, suddenly look as diverse as they were all along.' Reflecting widespread disillusion with pan-Arabism, he concluded that 'the pan-Arab idea that dominated the political consciousness of modern Arabs has become a hollow claim'. Now, political thinkers are more willing to defy 'the sacred myth' and to 'acknowledge that the Arab states have gone their separate ways'. By no means all Arabs would agree with Ajami's thesis about the demise of pan-Arabism. Nevertheless, there is strong empirical evidence that loyalties to the individual states have taken root while pan-Arab unification has lost some of its appeal (Reiser, 1983). In a sense, one can speak of two coexisting national identities: a supranational, pan-Arab one and a state-based quasinational one. In Arabic, there are actually two words for nationalism: qawmiyyah is commonly but not exclusively used to denote Arab nationalism, whereas wataniyyah is usually, but again not exclusively, used to refer to state-level attachments (Hudson, 1977, pp. 36-37).

The Syrian case

Syria illustrates the complex relationship between the two types of national identity and the way in which a state-based one has evolved. At its birth, Syria had no satisfactory, indigenous state-idea, having been created by the French after World War I. As a colonial artifact, it originated through no felt need by those who lived in it and its hastily drawn boundaries ignored fundamental cultural, historical, economic, and geographic relationships within the Levant. To the degree that its inhabitants identified with an entity called Syria, it was an ill-defined space, bilad al-Sham, which included the newly created Palestine, Lebanon, and Jordan. However, this Greater Syria was generally conceived of as a region within the Arab homeland, not a discrete political entity.

Syrians long viewed their state as an extemporaneous rump. Ajami remarks on the 'national consensus of sorts' that 'Syria's mission transcends her boundaries and that within those boundaries she is an amputated state' (1981, p. 124). Thus, from its inception Syria had a fundamental political geographic flaw. Many of its political leaders felt obliged to reject its very existence since to do otherwise would be to acquiesce to an imperalist fait accompli and accept the partition of the Arab world. One president in the mid-1950s disparagingly referred to Syria as 'the current official name for that country which lies within the artificial boundaries drawn up by imperialism when it still had the power to write history' (Seale, 1965, p. 130). Many of Syria's early post-independence rulers went out of their way to avoid the appearance of accepting the political geographic status quo and were reluctant to cultivate a specifically Syrian centripetal iconography. This is exemplified in a communique issued after one of many coups d'etat:

> Arab Syria and its people have never recognized the boundaries of its country and only acknowledge the frontiers of the greater Arab homeland. Even Syria's national anthem does not contain the word Syria, but glorifies Arabism and the heroic war of all the Arabs (Be'eri, 1970, p. 151).

Although Syrian wataniyyah is far better developed today than two or three decades ago, there is still some ambivalence about the state's identity. Syrian leaders, when they mention the country, often still refer to it as a region (qutr). The preamble to the constitution states that the 'Syrian Arab region' (emphasis added) is a part of the Arab homeland and that its people 'are part of the Arab nation which strives ... for ... complete unity'. In the past, especially, those who appeared to put the interests of Syria, or any other state, above those of the Arab nation were accused of regionalism (iqlimiyyah). The most important political body in the country today is the so-called Regional (i.e. Syrian) Command of the Ba'th party, which is theoretically subordinate to a pan-Arab National Command. Given the logic of Arab unity, the use of 'regionalism' (or 'separatism') was, in a sense, quite appropriate in this context.

Syria has always conceived of itself as the 'beating heart' of Arabism and as the guardian of Arab nationalist ideals. Arab nationalism, as an idea, first took root in Syria. Syria has also been at the forefront of most unity schemes and traditionally saw itself as the nucleus of a larger Arab state. Hinnebusch observes that it 'has been the strongest and most consistent centre of pan-Arab sentiment' (1984, p. 289). Hudson argues that 'the diffuse idea of pan-Arabism has remained perhaps the most widely and intensely held symbol of political identification in Syria since independence' (1977, p. 257). Nor is it by chance that the Ba'th party, which was founded specifically to bring about Arab unity, was born in Syria in the 1940s and has been in power there continuously since 1963. In fact, the only other Arab country in which the Ba'th has had any success at all is Iraq, where the party has ruled since 1968. (Ironically, Syria and Iraq, whose flags are almost indistinguishable, are bitter foes.) The Ba'th viewed the state-system as a 'mutilation' of the Arab nation. Its 'eternal mission' was to overcome this dismemberment; in its scheme, Syria was not a national unit but a base for this national mission (Hinnebusch 1984, p. 289). Or, as President Asad said recently in a moment of enthusiasm, 'We want Syria to grow larger in order to include the entire Arab homeland' (FBIS, 13 March 1985). Because of the allure and resilience of pan-Arabist ideas, loyalty to Syria as an independent political entity was long equated with betrayal of the pan-Arab suprastate-idea.

During the 1950s the Syrian state-idea was so weak, and the pan-Arab suprastate-idea so strong, that Syria's survival as an independent entity seemed doubtful. In 1958, in the midst of a crisis, Syria's leaders elected to erase Syria from the political map altogether, merging it with Egypt to form the United Arab Republic. However, the union was a dismal failure. Many Syrians deeply resented Egyptian domination. Disillusioned, Syria seceded in 1961 (Kerr, 1970). Ironically, the union was a turning point of sorts, doing much to fix the idea of Syria in Syrian minds. The whole experience helped to make Syria more aware of its uniqueness and of the practical difficulties involved in combining Arab countries. Syria has never again surrendered its sovereignty in the interests of Arab unity. Nevertheless, it entered into loose unity schemes with Egypt and Iraq (1963), Egypt, Sudan, and Libya (1969), Iraq (1978), and Libya (1980). The Ba'thi regime continues to exploit pan-Arabism as a legitimating ideology and clings to the fiction that Syria's interests are subordinate to those of the Arab nation as a whole.

Syria has, with the passage of time, gained a spatial coherence that it previously lacked, for reasons to be discussed later. Nevertheless, Syrians still have a somewhat blurred view of what Syria means or encompasses. President Asad expressed a sentiment shared by virtually all Syrians when he said, during a 1980 speech:

> In the recent past Arab Syria extended from Sinai to the Taurus Mountains. Who divided this Syria? Where is this Syria now? Why did they dismember Syria? Reaction, allied with colonialism, did all of this (FBIS, 24 March, 1980).

Neighbours, who do not share this sense of injury and fear Syria's power, worry about irredentist efforts to create a Greater Syria under Syrian domination. President Asad speaks for many Syrians (but no longer many Lebanese) when he claims that Syrians and Lebanese:

> are one single people, one single nation. We may be divided into two independent states, but that does not mean we are two separate nations ... The feeling of kinship ... runs deeper than it does between states in the United States (New York Times, 4 December, 1983).

Similarly, it is not unusual to hear Asad making remarks like 'we and Jordan are one country, one people, and one thing' (FBIS, 26 March, 1981) or to read newspaper editorials asserting that Jordan 'is a natural part of Syria. History has never recognized the presence of an international, or even administrative, entity separate from Syria' (FBIS, 27 April, 1981). Palestine, too, is often described as a part of southern Syria. For example, Asad once reportedly told Yasir Arafat, 'There is no Palestinian entity. There is Syria. You [Palestinians] are an integral part of the Syrian people. Palestine is an integral part of Syria' (The Nation, 1 October, 1983). A typical Syrian editorial admonishes: 'Remember, from the viewpoint of history, geography and struggle, Palestine is southern Syria and Palestine is the two banks of the River Jordan' (FBIS, 6 August, 1985).

National Unity

If one characteristic of many Arab states has been the iteration and elevation of supranational allegiances at the expense of national, state-based ones, another, paradoxically, has been the strength of subnational (i.e. substate) loyalties. Almost all Arab countries are stratified linguistically, religiously, or racially to some degree and this has often been an impediment to their unity.

Cultural divisions within Arab countries are complex and their salience is highly variable spatially. Generally, there are four types of state. First are those that are religiously diverse but linguistically cohesive: Egypt, with its simple interfaith division between a Sunni majority and Coptic Christian minority; Saudi Arabia, Yemen, and most of the Gulf shaykhdoms, with their simple intrafaith division between Sunni and Shi'i Muslims; and Lebanon and Syria, with their multiple intrafaith and interfaith divisions. Second are those that are linguistically diverse but religiously cohesive: Algeria and Morocco, which are Sunni Muslim but divided between an Arab majority and Berber minority. Third are those that are both religiously and linguistically diverse: Iraq, which has crosscutting Arab-Kurdish and Sunni-Shi'i cleavages, and Sudan, where religious, linguistic, and racial categories overlap. Finally are those that are divided between natives and non-natives: Jordan, where Palestinian refugees make up most of the population; and most of the oil producing states, where immigrants dominate the labour force (Drysdale and Blake 1985, pp. 150-170; Hourani, 1947).

There is no simple, direct relationship between where a country falls in this typology and the extent of its integrative problems. What matters ultimately is not the type of cleavage but the way in which groups interact, which in turn is a function of the distribution of power, the degree to which vertical and horizontal cleavages overlap, and the political and geographical coherence of substate groups, among other things. Descriptions of cultural patterns within a state often exaggerate differences by implying the existence of discrete polar categories. Not only does this invite reductionist explanations of unity problems, it ignores other, equally salient markers of social identity that may cut across and soften or negate vertical divisions. An individual's ascriptive and non-ascriptive cultural and social identities are numerous, multifaceted, interrelated, and often ambiguous. Whether these identities are ones of ethnic or sectarian affiliation, generational cohort, class, or profession, they are not necessarily mutually exclusive and they can each constitute a focus for group solidarity. Moreover, the relevance of these identities is situational and can be extremely subtle. Ethnic or sectarian identities may be pertinent in some contexts and not at all in others. The relative importance of a particular identity may change through time, and not necessarily unidirectionally. The fact that ethnic or sectarian identifications are salient or, equally important, irrelevant in a particular situation does not mean they will always be so. Traditionally, it was assumed that the growth of a capitalist economy, moderniza- tion, and increased spatial and social interaction in the Third World would undermine subnational allegiances and replace them with secular, class identifications and a single national identity. This view was rooted in the belief that subnational identities were ephemeral vestiges of a traditional way of life and incompatible with moder- nity. In reality, modernization supplements rather than supplants parochial identities with national ones. These perspectives are essential if one is to understand why not every minority group in the Middle East is agitating for autonomy or independence. Diversity should not be equated automatically with disunity.

Syrian unity

Syria illustrates many of these points and emphasizes the complex interplay between subnational, national, and supranational identities. Syria's population is overwhelm- ingly Arabic-speaking. Its Kurdish, Armenian, Circassian, and Turkoman language minorities are small and bilingual and do not present an impediment to national unity. Syria is, however, religiously heterogeneous. It must be emphasized that in the Middle East sectarian differences have an importance they often lack elsewhere in the world. Relig- ious identities are at least as important as linguistic ones, and possibly more so, in shaping subnational and even national political allegiances (the Iraqi Kurds being an obvious exception). In the Ottoman Empire, the principal distinction until the late nineteenth century was not between Turks, Arabs, Kurds, and other language groups but between Sunni and Shi'i Muslims and the numerous small non-Muslim or splinter Shi'i sects, most of which were granted autonomy to administer their own affairs under the millet system. For many centuries, religion, not language, defined an individual's fundamental identity. With the spread of language-based nationalism to the region after the late nineteenth century, awareness of ethnic distinctions supplemented rather than supplanted sectarian identifica- tions.

Syria's population is approximately 85 per cent Muslim and 15 per cent Christian. The Christians were prominent in the early Arab nationalist movement and were among the first proponents of secularization (which would obviously be to their benefit in an overwhelmingly Muslim society). Unlike Egypt's Christians, almost all of whom are Coptic, and Lebanon's Maronites, whose protonationalism has contributed to that country's fragmentation, Syria's Christians are subdivided into numerous small sects, have no separatist agenda, and no large rural territorial base.

The Muslims, like the Christians, are subdivided. Most are Sunni, but some 15 per cent of the population is Alawi and 3 per cent is Druze. The beliefs of these splinter Shi'i sects are far enough removed from mainstream Islam so that many Muslims do not consider them to be Muslim at all. Historically, Christians and Jews were tolerated as ahl al-Kitab, people of the Book, but the Alawis and Druzes were persecuted because the Sunni majority considered their doctrines heretical. Their survival can be attributed in part to the fact that, much like three other major minorities in the region, the Kurds, Berbers, and Maronites, they inhabit inaccessible, peripheral mountain zones, which have afforded them some protection. The Alawis are heavily concentrated in the mountains of northwest Syria, a particularly impoverished region until recently, while the Druzes are almost all located in a rugged upland region in the southwest. Both groups enjoyed a large measure of independence until well into this century. Moreover, the French, during their colonial mandate between the two world wars, created separate ministates for the Alawis and Druzes, more because they sought to rule Syria by dividing it than because they had a particular sympathy for these minorities.

Syria exhibits a common pattern, with vertical sectarian and geographical cleavages coinciding with horizontal class ones to some degree. On the basis of national integration experiences elsewhere, one might be tempted to predict the outcome: the Sunni majority dominates the state, while the sectarian minorities, their communal identities reinforced by geographic isolation, political oppression, religious persecution, economic exploitation, and social discrimination, struggle for autonomy within or independence from a state with which they do not and cannot identify. But this is not at all what happened. Sunnis did, indeed, dominate the state until 1963. In addition, Alawis and Druzes unsuccessfully resisted attempts to undermine their autonomy during the 1950s as the power of the political and bureaucratic centre grew and a highly centralized unitary state emerged. But in almost every other respect what occurred was unusual.

In Syria the majority is in a subordinate position and real power resides where one would least expect to find it: with the community that traditionally has been most disadvantaged. The Alawis, Syria's largest minority, have played a pivotal role in the country's political life since 1963, largely because they have been overrepresented in the armed forces and the Ba'th party (van Dam, 1981; Rabinovich, 1972; Drysdale, 1979, 1981b, 1982).

During the colonial period, the French recruited a disproportionate number of soldiers for their local forces, the predecessor of the present armed forces, from the minority communities. In part this reflected France's divide and rule strategy: the nationalist movement was heavily Sunni Arab and the French believed they could use the minorities, whom they considered to be more reliable politically, as a counterweight. In addition, the French had crude notions about the martial attributes of such groups as the Alawis, Druzes, and Kurds, portraying them as 'born soldiers'. After independence in 1946, an effort was made to correct these imbalances. Nevertheless; Alawis, especially, continued to enlist in large numbers, particularly in nonofficer ranks, because the military was one of the few avenues of upward mobility and one of the only national institutions where their sectarian affiliations did not put them at a disadvantage. The Alawis were attracted to the army precisely because it was a modern bureaucratic organization in which rank was a function of performance and ability, not class and primordial identity. Military service also had an obvious appeal to a people who lived in the most impoverished part of the country and who formed an exploited underclass. Military service became a strong Alawi tradition and an important source of income in the poor mountain villages.

Alawis were also attracted in disproportionate numbers to the Ba'th party, which promised to create a secular, egalitarian Syria in which they (and other disadvantaged minorities) were not automatically consigned to the bottom of the ladder because of their sectarian identities. During the 1950s and early 1960s, the Ba'th gained support as the most dynamic party in Syria. Its platform of Arab unity, land reform and socialism was popular especially in rural areas and among young officers.

In 1963, a group of Ba'thi officers seized power. Virtually all of the leading participants in the coup either came from the minority communities, particularly the Alawi one, or were Sunnis from rural areas or from small provincial towns in peripheral provinces. Almost all were young and had peasant or petit bourgeois origins. As a result, there was an historic shift of power away from the traditional Sunni ruling class of Damascene and Aleppan merchants and landowners to the rural periphery (Van Dusen, 1975). Many Sunnis deeply resented the reversal and viewed the coup as purely sectarian and revanchist. Although sectarian ties unquestionably lubricated the Ba'th's climb to power, it would be simplistic to depict the coup in solely sectarian terms. Sectarian, rural, class and socialist identities were closely interwoven. This was well illustrated in a remark attributed to a Syrian officer after the Ba'thi coup d'état:

> Don't expect us to eliminate socialism in Syria, for the real meaning of such steps would be the transfer of all the political, financial, industrial, and commercial advantages to the towns, i.e. the members of the Sunni community. We, the Alawis ... will then again be the poor and the servants. We shall not abandon socialism, because it enables us to impoverish the townspeople and to equalize their standard of life to that of the villages (Be'eri, 1970, p. 337).

Whatever the motives of those who seized power, the perception that the regime had exploited sectarian, tribal, and kinship ties to consolidate its position and had a distinct minority colouration presented the Ba'th with a serious legitimacy problem. Sunni apprehensions grew as the armed forces were purged and Alawis promoted to many of the key upper echelon positions in the officer corps, all under the pretext of creating a loyal 'ideological' army. The more vulnerable the regime felt, the more it relied on sectarian ties in self-defence. The irony of a secularist political party behaving in this way did not escape attention. All the while the regime inveighed against ta'ifiyyah, or sectarianism.

Alawis have never acted in unison. Since 1963, the most bitter power struggles within the regime have been waged among Alawis, who are tribally and ideologically divided. Factional divisions, which culminated in intra-Ba'thi coups d'état in 1966 and 1970, have seldom coincided completely with sectarian ones. Nonetheless, since 1963 Alawis have, in effect, controlled Syria. President Asad is an Alawi, as are many of the key commanding officers (Batatu, 1981, pp. 331-332). The regime has looked increasingly like a narrowly-based sectarian one, although Sunnis have never been totally excluded from power and traditionally occupy some of the most visible positions, like prime minister. In the mid-1970s, opponents began a wave of bombings and assassinations of Alawis. The regime responded with repression, which further alienated many sectors of the population. Muslim fundamentalists have played a leading role in opposing the regime, which they regard as being both Godless because of its professed secularism and illegitimate because it is controlled by heretics. In 1980, large-scale anti-Alawi and anti-regime disturbances broke out in Aleppo and Hamah. Two years later, a popular uprising against the regime in Hamah was brutally squashed. The city was virtually levelled and some 10,000 civilians killed by a mostly Alawi praetorian guard (Drysdale, 1982b). Sectarianism has deeply infected Syrian life and many worry that there will be an Alawi bloodbath after the regime is finally overturned and the Sunni majority assumes power again.

If one of the many ironies of Syria's situation is that an ostensibly secular, socialist political party has exacerbated sectarian relations, another is that under the Ba'th, which was founded in order to bring about Arab unity, Syria has shied away from the promiscuous unionism of the early postindependence era and has acquired the distinctive 'national' identity it historically lacked (Ma'oz, 1972). In part, this reflects a general trend throughout the Arab world. The failure of so many Arab unity schemes has had a sobering effect. Arabs are far readier to admit that there are real differences among the states that cannot be wished away. The interests of the individual Arab states clearly take precedence over pan-Arab interests and Syria is no exception, official ideology notwithstanding. Everywhere, it is now accepted that loyalty to the individual states can coexist with pan-Arab ideals, especially in their most abstract form. Syria has existed for well over half a century and Syrians have naturally come to identify with it and to think of themselves as Syrian as well as Arab. What seemed artificial at birth has acquired an air of permanence, if only because a tiny minority of Syrians can remember a time when the state did not exist. States have a way of acquiring their own justification.

In other respects, the development of a specifically Syrian national identity can be attributed to the Ba'th. The Ba'thi regime's deficiencies - its repressiveness, its narrow sectarian base, and its corruption and nepotism - are all too apparent. But it can also claim accomplishments in the area of state-building. Under the Ba'th, Syria has enjoyed relative stability for an unprecedented twenty-four years. This continuity has allowed the creation of a strong political centre, the introduction of ambitious national planning, and massive infrastructural development, which have encouraged the development for the first time of a national space economy. Since 1963, Syria's constituent parts have become more fully integrated and the economy has gained a functional coherence it formerly lacked. Under the Ba'th, special efforts have also been made to reduce regional inequalities, especially urban-rural and centre-periphery ones (Drysdale, 1981a). In addition, Syria has emerged as a major power within the region, especially since President Asad came to power in 1970. Although many Syrians oppose aspects of the regime's foreign policy, such as the involvement in Lebanon and support for non-Arab Iran in its war with Iraq, they nevertheless feel a sense of pride about Syria's military and political stature.

Another intriguing hypothesis is that Syrian 'separatism' is partly the result of minority control. Traditionally, Syria's minorities have both supported secularist political parties in disproportionate numbers and been less receptive to the idea of Arab unity, which would further dilute their influence. Whereas the Sunnis initially regarded the state as illegitimate, many from the minorities considered Syria's continued existence preferable to its incorporation within a larger Arab state. Thus, paradoxically they had a greater stake than the Sunni majority in Syria's perpetuation and independence. Before it was outlawed, the Syrian Social National Party, which was committed to creating a Greater Syria and rejected the pan-Arab state-idea, had an almost exclusively minority constituency. Allegedly, Alawi and Druze members of the Ba'th from its inception gave less priority to its pan-Arabism and more to its socialism. It would be wrong to suggest that Alawis and other minorities conspired to seize power in order to prevent Syria's unification with other Arab countries. Nevertheless, since 1963 a 'Syria first' policy has been much more in evidence, even while lip-service continues to be paid to pan-Arab ideals. As Ajami (1981, p. 124) notes, Asad is 'the first leader in modern Syrian history to make peace with Syria's national situation'. As a member of a minority, 'he has harbored no illusions about Arab unity'.

Conclusions

The national integrative experiences of the Arab states have been varied. Until relatively recently, they have generally been coloured by the strength of supranational and subnational allegiances and the corresponding weakness of state-based ones. In the past, national unity was understood by many Arabs to mean the reunification of the Arab nation. To put the interests of the states themselves above those of the Arab nation as a whole was, in a sense, to accept its partition, to foster subnational allegiances, and to engage in separatist activity. In the Syrian case, the illegitimacy of the state was profoundly debilitating initially. Today, few believe there will ever be a single Arab nation-state. Consequently, the states now view themselves in a rather conventional way and the identities they are cultivating are quasinational ones. Nevertheless, terms like nation, nationalism, and national identity still have several layers of meaning and cannot be used in the context of the Arab world without some qualification and refinement.

Majority-minority relations in Syria are, in certain respects, not typical of those elsewhere in the region. Nevertheless, they suggest that the conventional political geographic paradigm, in which the centrifugalism of autonomist or separatist peripheral disadvantaged minorities is countered by the centripetalism of a majority-derived state-idea, may not always be valid. Perhaps political geographers have paid too much attention to majority-minority relations in states which fit the European pattern. Examining the numerous cases in which minorities have seized power - hijacked the state - in the Third World or have played a leading role in state-building might prove fruitful in giving us a richer understanding of national integration dynamics.

References

Ajami, F. (1978-79). 'The End of Pan-Arabism', Foreign Affairs, 57, 355-373

Ajami, F. (1981). The Arab Predicament: Arab Political Thought and Practice Since 1967. Cambridge University Press, Cambridge

Batatu, H. (1981). 'Some Observations on the Social Roots of Syria's Ruling Military Group and the Causes for its Dominance', Middle East Journal, 35, 331-332

Be'eri, E. (1970). Army Officers in Arab Politics and Society, Praeger, New York

Clarke, C., D. Ley and C. Peach (eds.) (1984). Geography and Ethnic Pluralism, George Allen & Unwin, London

Drysdale, A. (1979). 'Ethnicity in the Syrian Officer Corps: A Conceptualization', Civilisations, 29, 359-374

Drysdale, A. (1981a). 'The Regional Equalization of Education and Health Care in Syria Since the Ba'thi Revolution', International Journal of Middle East Studies 13, 93-111

Drysdale, A. (1981b). 'The Syrian Political Elite, 1967-76: A Spatial and Social Analysis', Middle Eastern Studies 17, 1-30

Drysdale, A. (1982a). 'The Syrian Armed Forces in National Politics: The Role of the Ethnic and Geographic Periphery', in R. Kolkowicz and A. Korbonski (eds.) Soldiers, Peasants and Bureaucrats. Allen and Unwin, London

Drysdale, A. (1982b). 'The Asad Regime and its Troubles', Middle East Research and Information project (MERIP) Reports 12 no 8, 3-11

Drysdale, A. and G.H. Blake (1985). The Middle East and North Africa: A Political Geography, Oxford University Press, New York

Farah, T.E. (ed.) (1986). Pan-Arabism and Arab Nationalism: The Continuing Debate, Westview Press, Boulder

Foreign Broadcast Information Service (FBIS). Middle East and Africa: Daily Report

Haim, S. (ed.) (1964). Arab Nationalism: An Anthology, University of California Press, Berkeley

Hinnebusch, R.A. (1984). 'Revisionist Dreams, Realist Strategies: The Foreign Policy of Syria', in B. Korany and A.E. Hillal Dessouki (eds.) The Foreign Policies of Arab States, Westview Press, Boulder

Hourani, A. (1947). Minorities in the Arab World, Oxford University Press, London

Hudson, M. (1977). Arab Politics: The Search for Legitimacy, Yale University Press, New Haven

Kerr, M.H. (1970). The Arab Cold War, Oxford University Press, London

Kliot, N. and S. Waterman (eds.) (1983). Pluralism and Political Geography: People, Territory and State, Croom Helm, London

Lewis, B. (1964). The Middle East and the West, Harper & Row, New York

Ma'oz, M. (1972). 'Attempts at Creating a Political Community in Modern Syria', Middle East Journal 36, 389-404

Orridge, A.W. (1982). 'Separatist and Autonomist Nationalisms: The Structure of Regional Loyalties in the Modern State', in C. Williams (ed.) National Separatism, University of Wales Press, Cardiff

Paddison, R. (1983). The Fragmented State: The Political Geography of Power, St. Martin's Press, New York

Rabinovich, I. (1972). Syria Under the Ba'th: The Army-Party Symbiosis, Israel Universities Press, Jerusalem

Reiser, S. (1983). 'Pan-Arabism Revisited', Middle East Journal 37, 218-233

Rodinson, M. (1981). The Arabs, University of Chicago Press, Chicago

Seale, P. (1965). The Struggle for Syria: A Study in Postwar Arab Politics, 1945-58, Oxford University Press, London

Taylor, P.J. (1985). Political Geography: World-Economy, Nation-State and Locality, Longman, London and New York

van Dam, N. (1981). The Struggle for Power in Syria: Sectarianism, Regionalism, and Tribalism in Politics, 1961-1980, St. Martin's Press, London and New York

Van Dusen, M.H. (1975). 'Syria: Downfall of a Traditional Elite', in F. Tachau (ed.) Political Elites and Political Development in the Middle East, Schenkman Publishing Co., Cambridge, Mass.

Williams, C. (ed.) (1982). National Separatism, University of Wales Press, Cardiff.

Chapter 7

PROBLEMS IN COMBINING LABOUR AND NATIONALIST POLITICS: IRISH
NATIONALISTS IN NORTHERN IRELAND

Clive Hedges

It is not surprising to find regional or sub-state nationalisms in countries like France, Spain and the United Kingdom
that have in effect long been multi-national states. The
crucial cultural underpinnings of nationalist ideology have
not ceased to exist in areas such as Wales, Catalonia and
Brittany although these regional cultures in themselves are
no explanation for sub-state nationalism's renewed political
relevance in Western Europe over the last twenty-five years.
What is surprising is that an ideology generally considered
by political analysts to be more useful, in a European
context, for fascists or ailing right-wing governments
should have become linked with various forms of socialism,
as it has been in many sub-state nationalist movements in
Europe since 1960.

Flemish nationalism is a reminder that not all such
current European movements share a concern with socialism or
are free from fascistic tendencies but certainly most of
those areas with substantial and lasting electoral support
for sub-state nationalism such as Ireland, Wales, Scotland,
Catalonia and the Basque Country all have political parties
who declare themselves to be socialist in some form or
another, even if they may be in competition with conservative nationalist parties as in the last example. This
'leftist' orientation is not a totally new phenomenon: the
Irish Republican Congress was an attempt to forge links
between nationalism and socialism in the 1930s. However it
was a fairly isolated example whereas today such movements
are relatively widespread and play a more prominent part in
the political arena. It is the generic factors of nationalism that make this fusion with some kind of socialism so
interesting. Nationalism essentially claims to represent
the interests of the whole national population of a particular territory, the ethnic boundaries of the former and
geographic boundaries of the latter being defined by the
nationalists. It claims then to unify this population
internally, as Anderson (1986) puts it: in the past it has

generally denied or at least tried to ignore or play down the existence of class divisions in the interests of creating a unified nation. How then can such an ideology be reconciled with a programme of revolution or reform that claims to represent the interests of the working-class as a separate class, namely socialism?

Logic would indeed suggest the two were incompatible but, as indicated, many sub-state nationalist political parties have nevertheless attempted or claimed to have fused them; so how and why are they doing this? The answers to both questions are essentially political as although cultural traditions are crucial in defining national identity, the national interest of this entity and thus the definition of the nation is a political issue. It is not surprising that such self-conscious ideological reorientation has led to numerous divisions and splits within nationalist movements and parties.

How the nation is defined is central to nationalism as an ideology but such definitions are not arrived at in a social vacuum. Nationalism, like other ideologies, can vary in strength from area to area within the designated national territory 'depending on its ability to tap a responsive chord among local populations', as Agnew (1986) puts it.

Although the splits and divisions within these movements can be on a number of issues, for example the use of violence, these issues and splits may be linked to variations in the social and spatial distribution of the supporters of sub-state nationalist movements. Socialism is an ideology generally based in the urban working class. Does this reorientation of some sub-state nationalist movements towards socialism indicate a growing number of urban working-class supporters or at least a desire for such on the part of the nationalist? Furthermore what does socialism mean for these movements and how do they reconcile it with nationalism?

The Case of Northern Ireland

The character of Irish nationalism, and particularly its most militant form, Irish republicanism, has undergone a number of dramatic changes since the 1960s and like many other separatist and irredentist movements has become increasingly concerned with social and political issues to the extent of claiming to be socialist. The two main nationalist parties in the North of Ireland are Sinn Fein and Social Democratic Labour Party (SDLP). The latter party believes in the possibility of reforming Northern Ireland as well as in the possibility of unifying Ireland some time in the future. It characterises itself as 'left of centre' in its constitution. Sinn Fein claims to be a revolutionary, socialist republican party and supports the armed struggle of the Irish Republican Army for a united Ireland. There has been a number of splits in both parties in recent years. Gerry Fitt resigned from the leadership of the SDLP declaring the party to be increasingly more nationalist and less socialist. Sinn Fein split into two parties in 1970 over the issue of abstentionism from the established parliaments in Ireland and the United Kingdom. The roots of this split, however, lay in attempts to come to terms with socialism. Most of·those who disagreed with the policy of abstentionism have since gone on to form the Worker's Party who disavow violence and subordinate the national problem to the need for working-class solidarity.

Surveying the ideological output of nationalist parties is one way to examine these questions, and the way most frequently used, combined with a number of interviews with people involved in the two major nationalist political parties in Northern Ireland. Whereas large scale political surveys of the general population in an area like Northern Ireland are very difficult to carry out with any hope of a large response, it is relatively easy to gain access to all but the most prominent politicians. Interviews with leading members of each party, Austin Currie of the SDLP and Danny Morrison of Sinn Fein, were carried out as well as three interviews with local party activists or councillors in each party, firstly in Belfast and secondly in the county of Fermanagh, for reasons outlined below.

The central geographical division in nationalist politics in the North of Ireland is that between town and country. Between World War II and the 1960s Catholic politics were dominated by the Nationalist Party which didn't have any formal structure but was rather a loose aggregate of small farming and business interests usually overseen by the Roman Catholic Church. Nationalists were still heavily discriminated against in these years in terms of political representation but the narrow conservative, petit bourgeois and religious interests that made up this party were incapable of mounting any real opposition to Unionism.

In Belfast the situation was different to the degree that although political participation wasn't high those candidates that were returned were generally of a republican socialist nature. Their problem was that their version of socialism was one which was centred around trade-union work and the problems of the urban population: they paid little attention to the position of small farmers, for example, and so generally restricted their appeal to an urbanised working-class whilst their nationalism varied from a staunch republican position to very mild nationalist sentiments, thus giving enormous room for splits on the national issue. In other words, there was never any real attempt to link town and country nor was there any coherent position on the relationship between nationalism and socialism. Sinn Fein and the SDLP, in attempting to forge an ideology that will be unifying across the province, and for Sinn Fein across Ireland, have had to tackle both of these problems. Nevertheless this division still remains in similar form. It is a division between the East and the West of the province as well as between town and country however.

It is the counties in the west of the province, particularly Fermanagh and Tyrone and, to a lesser extent, Armagh and Derry, that have always been core areas for the I.R.A. in terms of support and activities. It is this area where the old Nationalist Party had its greatest support and where the Irish Independence Party, the closest to a more modern equivalent there has been, gathered most of its support before Sinn Fein began to take elections seriously. It is where, as it were, the traditions of nationalism are most obvious.

The division can be best represented by the county of Fermanagh and by West Belfast, the former being an area of mostly relatively small farmers, the latter being the major urban centre of the nationalist population in the North. The former now has a Sinn Fein-controlled district council whilst the latter, although it has a Sinn Fein member of parliament, is now the main area of contention between the two nationalist parties. The differences between the two areas are such that they have a definite effect on the way in which the two parties have developed and these differences illustrate that as a form of practical politics nationalism, as already indicated, becomes vulnerable to spatial variations in its appeal, as Agnew (1986) has found with Scottish nationalism.

The effects are most marked within the SDLP who feel the area of Fermanagh is unsympathetic to their position, as Austin Currie put it:

> Fermanagh/South Tyrone and Mid-Ulster have always been considered to be the most nationalistic constituencies ... and the more remote the area is the more pro-Republican it is likely to be.

John O'Kane, a local SDLP councillor in Fermanagh, agreed:

> I would find that most Catholics who are supporters are really, here in these counties, even the SDLP supporters, are the old green nationalists ... their attitude would be 'keep the Unionists out'.

O'Kane sees local politics functioning on the basis of being anti-British, a similar type of nationalism to that of twenty or even sixty years ago. As Eugene Kelly, another SDLP member in Fermanagh, puts it, it is:

> a reactionary political choice ... the reaction of people living in the rural areas is that it's the only way we can show our opposition to Unionism ... and the dominance Unionism had in a county where there was always a nationalist majority.

The effect of this on politics of the local SDLP is to push it towards a more traditional nationalist position, in O'Kane's view creating an obvious split in the party:

> East of the Bann, the Belfast area, it would be more Alliance-type people, whereas west, Fermanagh, Tyrone, Derry, we have to be more pragmatic, we are effectively more green-nationalist.

For example Jim Lunny, an ex-member of the old Nationalist Party and now chair of Fermanagh District Council, is one of the few members of the SDLP in such a position who would call himself a republican and admit that: 'I joined the SDLP because it was the Irish party'. A more obvious example of the effects on local politics is the willingness of the SDLP councillors to co-operate with their Sinn Fein counterparts and vote, for example, for a Sinn Fein chair in 1985 despite what O'Kane calls 'a lot of pressure from Belfast (the SDLP headquarters) on our votes' to elect a Unionist chair, but:

> The feeling of the people who supported us was keep the Unionists out and it would have been suicide for us to have gone and voted for the Unionist side ...

This is quite a different view to that expressed in Belfast. The party members here are generally insistent that such a split does not exist within the party, admitting, as Albin Magistrate does, that 'mid-Ulster and Fermanagh/South Tyrone are tougher than West Belfast' for the SDLP but resenting the 'idea of their being different wings in the SDLP ... the SDLP is a fairly homogenous political entity, I do not believe it is a disparate party ... and I do not believe, in any substantive sense, people are more nationalist in some areas'. However, this denies the experience of the party in Fermanagh, particularly since the 1981 Hunger Strikes when it decided to stay out of the Fermanagh/South Tyrone by-elections. The main question, however, is not just whether one area is 'more' nationalist than another, but what form this nationalism takes and whether the appeal of nationalism is to a degree 'contextually constituted' as Agnew (1986) puts it.

Sinn Fein, although also keen to stress the unity of a province- and island-wide ideology, do recognise some differences between the two areas basically due to the differences between rural and urban areas. Danny Morrison of Sinn Fein states that:

> outside of those two large cities (Belfast and Derry) and perhaps towns like Newry and Armagh there wasn't necessarily great consciousness about being working-class citizens.

And besides this, 'socialism is ... largely originally an urban phenomenon'. This doesn't mean that the party was necessarily any different in different areas, although some did believe that it was more difficult to raise social problems as political issues in rural areas, but the purpose of the party is to 'represent our people' and 'quite a lot of our people especially in the six counties are farmers themselves and recognise the problems, in the same way we would represent people on the dole here (Belfast).', as Paddy Molloy, a Belfast Sinn Fein member, puts it. In other words the variations in nationalist consciousness do not present the same degree of difficulty for Sinn Fein in articulating an ideology, although Morrison admitted that their large rural support did obviously affect their position. There could be no serious involvement with a Marxist analysis, for example, because 'it is a very anti-rural point of view ... in terms of the nationalisation of land and the struggle of nationalist farmers' as he put it. Thus to the SDLP in Fermanagh Seamus Mallon's victory as an aggressive nationalist candidate in Armagh in early 1986 is indicative of the different nature of politics in the rural west. To the party in Belfast it was because 'he was an SDLP candidate' and to Sinn Fein it was 'because he was a nationalist ... and because it was winnable'.

How then can we see the effects of these spatial variations on the parties' political positions in general? The most damaging possible effect on the SDLP is on its claims to speak for the majority of the northern nationalist population and have a base throughout the province. These are vital components in its ability to win any effective reforms from the British state, and in its ability to garner support for those initiatives which are taken. It is possible for the SDLP to claim Seamus Mallon's electoral victory was a result of the Anglo-Irish agreement, for example, despite his relatively hard-line nationalist views, but it is not possible to ignore the uneven manner in which this agreement has been received by the nationalist population. Whereas in Belfast Maginness can claim the agreement to be:

> one of the most important events since partition ...
> it has attempted to establish equality between the two
> different traditions.

and Margaret Ritchie, an SDLP councillor in South Down, can
claim

> that one of the long-term consequences of the Hills-
> borough Agreement is that Unionists have had to face
> defeat.

the view from members in Fermanagh is decidedly less
optimistic. Eugene Kelly admits that:

> there isn't much to say at the minute from the point
> of view of a nationalist, you don't see any change.

and in O'Kane's view:

> it doesn't really promise all that much for national-
> ists, the things it does promise are small steps
> really ... compared to the removal of the UDR for
> example ... it might just peter out.

Sinn Fein is still in the process of altering its
ideological stance so as to combine traditional nationalist
aspirations about the armed struggle with a developing
concern for social issues and demands for electoral participation. Most historians and contemporary commentators on
the Provisional I.R.A. and Sinn Fein characterise them as
conservative reactionaries at the time of their inception
and yet still manage to see them as 'socialist' now. In
fact a thorough reading of their propaganda at the time of
the split in the republican movement shows that for every
mention of a communist plot to subvert republicanism there
is a plan of the new socialist republic to follow independence. The Provisionals in fact often saw the national
struggle as apolitical but themselves as socialists because
they saw no possibility of socialism being achieved outside
a united Ireland, a way of thinking with a historical
background in Belfast politics in particular and the
sectarian nature of the Northern state (see Munck and
Rolston, 1986). They have now arrived at a position
remarkably similar to the one they refused to endorse in
1970, however, when they split from the rest of the republican movement. It is an ideology that characterises the
political struggle as the building of a broad democratic
movement against imperialism, similar to that of the A.N.C.
for example. If any one series of events can be seen as
central to this it is the Hunger Strikes of 1981. As
Morrison put it:

> had the Hunger Strike not taken place ... it probably
> would have been very difficult for us to have taken
> the movement from Sinn Fein's role of just selling
> newspapers.

The rhetoric of Gerry Adams, head of Sinn Fein, was for the
first time concerned with forming broad political fronts on
humane grounds in order to win the demands of the Hunger
Strikers and this was a dramatic change of tactic from that
used in earlier Hunger Strikes in the 1970s. There was a
pressing need for the movement to make their struggle less
elitist, to react to the social problems of their community
which had been turned into political issues by the Civil
Rights movement and the SDLP, and to counter the mandate
given to the SDLP in elections to speak for the nationalist
issue. It was also becoming obvious that some military
tactics were becoming unpopular and that the IRA did not
have the necessary strength to force British withdrawal in
the near future. In other words to remain dynamic it had to
take part in practical politics as well as reproduce
nationalist traditions.

This has been attempted through the formation of an
anti-imperialist ideology. Thus in their newspaper there
are denunciations of American foreign policy and comparisons
between the IRA and 'Third World' movements, into which is
inserted the history of Irish armed struggle, so giving both
historical 'continuity' and present-day 'legitimisation' for
their position. The link between the armed struggle and
electoral politics then becomes 'socialism', which demands
the opposition to British imperialism and the political
representation of 'our people'. The publicisation of social
problems and the use of welfare or advice centres can then
be neatly dovetailed into the effects of British imperialism. In reality, of course, this is not so easily achieved
as any such form of practical politics is, as Morrison put
it:

> full of dangers of going too fast, of the leadership
> moving ahead of the grassroots, or of people mistaking
> a development for a reversal of strategy.

The current major problem for Sinn Fein has been how
to remove the policy of abstentionism in the Republic of
Ireland without alienating more traditional members and
supporters. It was in fact this change that finally split
the movement in 1969 and opposition to it has come from the
more 'peripheral' rural areas such as Galway and from
traditional nationalist members of the party such as Ruairi
O Bradaigh of Roscommon, once party leader, who has now led
for the second time led a walkout from a party conference

over this issue in order to set up a rival organisation opposed to such involvement in politics. Although it would seem Sinn Fein has achieved this manoeuvre it is still evident that in more rural western areas of Ireland as a whole, as well as Northern Ireland, a nationalism persists that stems from the experience of small farmers and is closer to its specifically Irish historical conditions. This ideological variation can limit change, as Paddy Molloy admits:

> especially in the rural areas in the 26 counties where you have a republican movement that has been an uncompromising organisation of ideas.

It is easier in some ways for Sinn Fein to accommodate narrow nationalism than it is for constitutional groups as long as it keeps up the armed struggle, thus keeping alive Irish traditions whilst justifying it to the more progressive elements by the use of anti-imperialist ideology, and broadening the base of support whilst still glorifying the nation and not 'selling out'. Molloy, again:

> there is still an element of people who want a united Ireland and that's it, so we have to say to ourselves the Republican movement is a broad-based movement and it will have to take in, by necessity, as many people as possible ... like the ANC in South Africa.

There has been then an obvious change in the perception of the relationship between nationalism and socialism. National reunification is obviously still the only aim of the armed struggle but no longer the only aim of the political struggle, as Molloy put it:

> once you get to that stage (of a united Ireland) you have to say, okay, we'll get off here, we're moving on ... and this is what we want.

Possibly the best example of this change is Sinn Fein's current view of the place of the Unionist population in Ireland. Traditionally, they have seen them as being duped by the British and failing to realise their real Irish identity; increasingly now, however, they speak of them, both worker and boss, as a colon class similar to South African whites, whose position is guaranteed by Britain:

> it [the sectarian divide] will take an awful long time to heal but undoubtedly it will. If Britain were to leave this island tomorrow the process would start then, with that factor here the working-class Union-

ists see themselves as slightly better-off than their nationalist counterparts.

Both political parties can in fact be seen as being in an ambiguous position towards the working-class as a class. Although they claim to be socialist they only strive to unite and represent one fraction of the working-class, those who belong to the nationalist community. Given the fundamental lack of congruence between those who consider themselves Irish nationalists and the working-class as a whole in Ireland, why try to reconcile nationalism with socialism at all? What are the events that have led to re-emergent Irish nationalism being reconstituted in the manner described?

Generalising Sub-state Nationalism

In order to analyse this process seriously it has to be recognised that there are similarities between sub-state nationalist movements in Western Europe and that a generalising approach can be adopted. Obviously each region has its own unique history but the changes in patterns of social organisation and economic organisation there are not just due to the uniqueness of that region but the interaction of wider national and international changes in organisation with each area's character and it is the effects of this interaction that regional political movements react with. Perhaps within a very wide view it is possible then to link regional trends to wider ones connected to wider spatial patterns of investment and divisions of labour. Social relations of production in any one area are likely to change due to changes in overall patterns of development, so a new spatial division of labour is likely to cause economic, political and social disruption in areas affected. This doesn't mean that one can read off political 'effects' from economic 'causes'. The character of this disruption depends largely on the character of the area, its cultural and political history as well as the character of new spatial structures.

Clearly, a nationalist response can only occur in an area where a national consciousness and tradition exists to some degree. It is also obvious, however, that nationalist sentiment does not automatically trigger a mass sub-state nationalist political response. In order to understand the whole process, one has to understand the nature of the disruption and analyse a nationalist response as a form of practical politics resulting from a reaction to contemporary forms of social organization.

The emergence of Basque nationalism was, for example, a reaction to industrialisation and a large influx of non-Basque labour into the area to fuel the growth of industrial capital. The change in spatial patterns of capital and labour in Spain led to a reaction by the local petit bourgeoisie against modernisation giving rise to a backward-looking, conservative ideology. In this form it was a coherent response to a change in the region's position in national and international structures of economic organization. This is obviously a simplified example, but illustrates a framework in which to understand these movements. Any explanation must be able to avoid both cultural and economic reductionism and perhaps this is best achieved, as Cooke (1984) says, by articulating the relationship between the cultural/ideological level and the level of production:

> in terms of the way class relations and social relations arranged upon other bases of collective identification, notably ethnicity, interact in civil society but are formed into instances of political mobilisation by active, conscious political organisation, including the organisation and reorganisation of practical ideologies.

The economic structures of Northern Ireland have undergone similar changes to other depressed areas of the United Kingdom and the failure of Premier O'Neill to relocate the region in international patterns of investment and retain the structures of political hegemony are well documented. The destabilising affect on Unionist politics is obvious but the changes in nationalist politics as shown are also very marked. The usual 'liberal' explanation is that changes in Catholic educational opportunities linked with non-discriminating externally-owned investment led to a rising Catholic middle-class that demanded a form of democracy. In fact, multinational capital invested in areas already developed by local and national capital and the rise in Catholic professional or managerial types was not as great as the rise in their representation in unskilled manual jobs. In as much as individuals are responsible for an upsurge in political activity then such changes in the Catholic social hierarchy did have an affect, a number of civil rights leaders being university students or teachers. However, it can't be doubted that the increasing sector of Catholic unskilled labour provided a discontented, anti-Unionist mass attracted by the issues involved and the forms of demonstration. In other words a hitherto inactive fraction of the working-class had now become important actors on the political stage. Irish sub-state nationalists, like those in Wales and the Basque Country, have since had to come to terms with this section of the population not just as part of the national mass but as part of an urban working-class. It is an indicator of how importantly both Sinn Fein and the SDLP treat their claims to be socialist and to represent the interests of the nationalist work- ing-class that they are often stressing the point that they do garner support from the working-class.

It is difficult to gauge accurately any particular differences between the support for one or other party in class terms and although election statistics can illustrate variations across space, variations between different strata of society within certain areas are more difficult to measure. This is further hampered by the fact that the spread of wealth among the catholic-nationalist population is still a fairly narrow one in terms of the population of Northern Ireland as a whole, partly due to discriminatory employment practices and policies. Sinn Fein like to distinguish their support from the SDLP as being more 'working-class' and including larger amounts of small farmers and the unemployed, according to Danny Morrison of Sinn Fein:

> our support tends to be working-class, in the rural areas the more poorer people than the well-off and middle-class Catholic types and a younger element.

According to Morrison again

> The Irish Times did an analysis of its [the SDLP's] councillors and they turned out to be doctors, teachers, solicitors, big farmers.

The SDLP are keen to deny any definition of themselves as 'middle-class'; they claim a much wider range of members, as Dan Keenan a Belfast SDLP member put it,

> I think SDLP membership to a certain extent reflects the Northern Ireland community, there are people who are very comfortably well off ... But I would say that there's quite a broad spectrum of membership in the SDLP and that wouldn't be the case in Sinn Fein.

However this allows the SDLP to present Sinn Fein's presence as the result of despair amongst the unemployed that can be soothed away by policies of reform, as well as to credit the existence of a Catholic middle-class partly to the reforms they have advocated in the past. The key area to hold, in order to claim to be speaking for the Nationalist work- ing-class, is the parliamentary seat of West Belfast currently held by Gerry Adams of Sinn Fein. However much they stress their socialist credentials, however, both parties are undoubtedly nationalist and as such their primary electoral aim is to achieve or maintain a position where they can claim to be speaking for the majority of the nationalist population and so also claim that they represent the nationalist desires of that population.

Thus it is perceived as totally consistent with 'socialism' to represent only those members of the working-class who belong to the nationalist community. In Sinn Fein's anti-imperialist ideology membership of this community is defined on political grounds as much as cultural ones, that is whether one is for or against a united Ireland. Those that ally their interests with the 'enemy', British imperialism, forfeit their position as members of 'the nation' and in effect become enemies of 'the people' or 'civilian hirelings' of the security forces and thus imperialism, as <u>An Phoblacht/Republican News</u> puts it (31.8.86). They thus arrive at the present policy of supporting the threatening of the lives of those workers who sell their labour to builders contracted by the police or British army. The contradiction then between the interests of the nation and the interests of the working-class as a class are easily ignored as long as the movement remains within communal boundaries and equates socialism with anti-imperialism and representation of 'our people'. The problems of uneven distribution are in effect blamed on Britain:

> The loyalists will only fight in proportion to the amount of hope Britain can give them for being able to hold on to sectarian privelege (Morrison in Collins ed. 1985).

Conclusions

Clearly whatever the 'objective' conditions in which such movements as the IRA operate, the formation of a coherent ideology based upon anti-imperialism is possible and it can gather electoral support. In order to understand such an ideology, one really needs to study the beginnings of 'national liberation ideology' in the works of Mao Tse-Tung, which first developed the formulations underlying much of the thinking of anti-imperialists and radical nationalism. Crucial to his tactics was the need to form class alliances between the proletariat and the peasantry and with sections of the bourgeoisie. It was important for him to be able to represent various class interests without apparent contradiction. During revolution this was done by characterising the enemy as feudal, imperialist and generally anti-progressive against which all 'progressive' elements should combine, as a result his rhetoric hailed the 'people's democratic dictatorship' rather than that of the proletariat. Resolution of any contradictions in the reconstituted

society was easy, said Mao, because 'the basic contradictions in socialist society are still those between the
relations of production and the productive forces' (in 'On
contradictions' in Schram, 1969) but in socialism these
contradictions are democratic and non-antagonistic. This
theoretical back-flipping is possible because of the concept
of 'the people', a near national body that encompasses the
working-class and their interests to a degree and so makes
it possible for nationalists to be allied to the proletariat
as a class rather than as merely part of the national mass.

It is these kinds of concepts and political formulations put into practice in various ways by a number of
'Third World' liberation movements that Sinn Fein has drawn
upon in building their own anti-imperialist ideology. In
some regions, of course, such a response is inappropriate,
although the position of E.T.A. in Spain is very similar. In
Wales, for example, the successful response for the nationalist Plaid Cymru has been a switch to labourist socialism.
The reasons for these changes lie not just in party political development but in the interaction in civil society of
local historical relations of gender, class, religion and,
of course, ethnicity with the social relations of production
that also vary over time and space, but each case has to be
studied in its own right, as well as part of wider patterns.

The failure of non-nationalist radical politics to
overcome sectarianism, the renewal of communal violence and
the presence of the British Army all added to the resurgence
of a nationalist response in Northern Ireland but in order
to last longer than the initial resumption of violence and
repression, it had to be a different form of nationalism and
it would be wrong to see these changes as purely cosmetic.
The problem was how to combine nationalist tradition and
contemporary issues and this has been attempted by the IRA
through anti-imperialist radicalism. The willingness of
numbers of feminists to agree to their position on the need
for national liberation in order to achieve women's liberation in return for a shift on issues such as abortion
encapsulates the possible way in which such an ideology can
succeed. However, it is clearly more than just a matter of
rearranging the rhetoric. In trying to unite the 'nation'
behind them sub-state nationalists have to face the problem
that the attraction of their nationalism may be radically
different in different areas of their 'nation'. The study
of Ireland gives one example of an attempt to overcome this
problem by the use of a form of 'socialism' based on an
anti-imperialist position.

References

Agnew, J. (1986) <u>Nationalism: Autonomous Force or Practical Politics</u>, paper presented to the Institute of British Geographers Annual Meeting.
An Phoblacht/Republican: <u>Official Organ of the Irish Republican Movement</u>.
Anderson, J. (1986) 'On Theories of Nationalism and the Size of States', <u>Antipode</u> 18, 218-32.
Berresford Ellis, P. (1973) <u>James Connolly - Selected Writings</u>. Penguin, London.
Bew, P., Gibbon, P. and Patterson, H. (1979) <u>The State in Northern Ireland 1921-72</u>, Manchester University Press, Manchester.
Collins, M. ed. (1985) <u>Ireland After Britain</u>. Pluto Press.
Cooke, P. (1984) 'Recent Theories of Political Regionalism: A Critique and an Alternative Proposal'. <u>International Journal of Urban and Regional Research</u>, 8, 549-72.
Massey, D. (1984) <u>Spatial Divisions of Labour: Social Structures and the Geography of Production</u>, Macmillan, London.
Minority Rights Group: (1982) <u>Report No. 9: The Basques and Catalans</u>, M.R.G., London.
Munck, R. (1986) <u>The Difficult Dialogue: Marxism and Nationalism</u>. Zed Press, London.
Munck, R. and Rolston, B. (1986) 'Irish Republicanism in the 1930s: New Uses for Oral History'. <u>International Journal of Oral History</u> 7,
Mao-Tse Tung: (1969) <u>On Contradictions</u> in: S.R. Schram: <u>The Political Thought of Mao-Tse Tung</u>. Penguin, London.

Chapter 8

SELF-DETERMINATION FOR INDIGENOUS PEOPLES:
THE CONTEXT FOR CHANGE

David B. Knight

Indigenous peoples in many states in recent decades have
experienced a rebirth of their distinctive senses of self
after decades, indeed centuries, of land dispossessions,
governmental denials, and a full range of societally
destructive forces. Many such peoples today find themselves
constantly threatened by dreadful oppression that is either
supported or simply ignored by governments; others live in
relatively open and accepting societies, yet they face
problems resulting from societal indifference, ignorance and
intolerance, and also from alien bureaucratic policies and
structures.

Indigenous peoples increasingly are seeking ways to
establish self-determination for themselves, but is self-
determination a reasonable, attainable goal for indigenous
peoples to strive for? What are the international prescriptions for the principle of self-determination? Do indigenous peoples qualify? What do indigenous peoples mean when
they call for self-determination? Are not indigenous
peoples simply sub-state minorities whose needs must be met
within existing states? Are there ways for indigenous
peoples to gain international recognition? These questions
have only begun to be examined in the political geography
literature (Knight, 1982, 1984, 1985a); they must be faced
head on as a prelude to the writing of a political geography
of indigenous peoples.

The purpose of this chapter is to tackle the fundamental issue of whether or not indigenous peoples have an
internationally accepted right to self-determination. By
clarifying this issue it may then be possible to provide a
template on which to place comparative studies so as to be
better able to determine the relative standing of indigenous
peoples within various states. In addressing this issue it
is necessary first to consider a variety of matters that
pertain to human rights, legal pronouncements, and the
behaviour of states, before then noting certain international actions and pronouncements by indigenous peoples, and

some contrasting ways in which the concept of self-determination is being defined by indigenous peoples. It is hoped that the interpretations and conclusions presented here will provoke response, for clearly a debate on the fundamental issues is called for. Any examination of self-determination necessarily draws from a wide range of interdisciplinary literature (Knight and Davies, 1987) although stress must be given to international law.

Dissenting Minorities, State Actions, and Human Rights

Dissenting minorities exist in many states. For some sub-state regional minorities the struggle against state oppression is constant, difficult, and, indeed, at times severe, for tyrannical governments (of the left and of the right) aggressively seek to control, subdue, and even eradicate those who do not conform to what is held to be the societal norm. Such governments may boldly proclaim a respect for human rights even as their actions make a sham of such proclamations. Other state governments may more fairly deal with dissenting minorities. Even in open liberal societies, however, actions against minorities, whether by implicit or explicit state action, constantly lead to watchful concern that stated guarantees for human rights are safeguarded. In such societies, the challenge may, for instance, find expression in the courts as people seek redress for past or present perceived or actual injustices.

Why are most, if not all, states open to charges of injustices, albeit to varying degrees? Governments generally seek to protect the status quo, according to varying ideological prescriptions and within certain traditions. Diversity, especially of opinion, is anathema to many contemporary governments. When the opinion expressed involves secessionist strands enunciated by leaders of a sub-state regional minority that finds itself quite apart from a national concept of belonging, then, perhaps not unnaturally, states will act to protect the status quo and seek better to manage, control or subjugate that minority. Integration and assimilation into the mainstream of national society often are to be found at the heart of state practices in dealings with dissident minorities. Means for encouraging or demanding integration and assimilation can be subtle or blatant, peaceful or violent. Equally, state management of diversity can be simple or complex, being based on traditional behavioural expectations or finding firm legal, constitutional, ideological, socio-economic, political and even military prescriptions. All states have limits to what is deemed to be acceptable behaviour. Who defines the behaviour? Ultimately it is the state, responding to the dominant societal demands. In short, the majority, which regards the state as theirs - or a minority which claims to speak for the majority - will prescribe limits for acceptable behaviours and will use the state as an instrument to protect those limits. But acceptability is in part a matter of perception. For instance, with respect to certain group behaviours, it will readily be recognized that separatists may, according to the state, be 'deviant' and thus may represent a threat that needs to be rebuffed, curtailed, or stamped out. In contrast, supporters of separatists may regard them as model citizens whose expectations and behaviours are to be emulated!

Most central governments declare that their 'subjugation' or 'curtailment' of dissension (by whatever means) is purely an intra-state matter. It can be argued, however, that questions of human rights and dissent within individual states must be linked to questions of human rights and dissent around the world. Accordingly, we must turn to international law.

International Instruments and Self-Determination

International law can be thought of as the codification of expected behaviours of states acting within an international system of states (Brierly 1963; Crawford 1979). International law can be rigid in definition but, because states' actions are involved, it can be quite fluid in practice. International law provides guidance to what is permissible and what is not with respect to claims for 'legitimate' self-determination. The United Nations Charter and many other U.N. documents include statements on self-determination. Thus, for example, it has been declared that:

> All people have the right to self-determination; by virtue of that right they may freely determine their political status and freely pursue their economic, social and cultural development (United Nations, 1960).

The U.N. pronouncements stress 'national unity' and 'territorial integrity', however, rather than a genuine concern for 'people' freed from a geographical (that is, state-wide) context. Indeed an overriding primacy is retained for the centuries-old foundations of international law and the system of states: recognition, state sovereignty and territorial integrity (Knight, 1984).

Two forms of self-determination have been recognized: internal self-determination, which refers to the choice a total people (or those who rule in their name) make when a form of government is selected; and external self-determina-

tion which refers to decolonization. I have argued elsewhere that new post-colonial forms of internal self-determination should be recognized, that is, where sub-state
regional identities are given recognition and political
control within their states with, in select instances,
limited forms of international recognition (Knight, 1985b).

Some departures from the colonial limitation are
permitted. First, self-determination is deemed to be
justifiable for populations ruled by a minority who govern
by an apartheid philosophy. Second, there is the acceptance
of a (very elusive) right to self-determination for people
of a state who are living under foreign domination. Third,
it may be possible for a sub-state regional minority to be
granted self-determination, but a major hurdle is presented,
namely, it must involve a free choice by the majority of the
total population. Relatedly, the U.N. has not yet recognized the concept of 'internal colonization'. With each
application of self-determination there is always the
fundamental issue of who the 'people' are.

The problem of definitions and interpretations

The U.N. Charter does not define 'people'. 'All people have
the right to self-determination ...' is the declaration, but
this has been taken to refer to the total population of a
state and not to 'minorities' within that territorially
defined 'people'. Thus the colonial application of self-
determination has been applied only to a total 'people' who,
at minimum, were said to be a 'people' because of being
located within set colonially-derived political boundaries,
no matter how culturally diverse the population was. It is
easy to appreciate why a different definition of 'people'
has been resisted by the U.N.: states wish it to remain
linked to the total populations of their respective states,
for self-determination can then be said to apply only to the
people of the totality of the existing territories.

The International Court of Justice and other bodies
have given some content to the term 'people'. For instance,
the ICJ jurists have proposed the following characteristics
of a people: (1) a common history; (2) racial or ethnic
ties; (3) cultural or linguistic ties; (4) religious or
ideological ties; (5) a common territory or geographical
location; (6) a common economic base; and (7) a sufficient
number of people (Indian Law Resource Center, 1984, p. 14).
These characteristics are similar to Shafer's (1972, pp.
17-20) ten basic characteristics of 'nationalism'. When
these concepts are linked to defining self-determination it
will be recognized that in trying to define terms there is
an inherent danger of being hopelessly caught in a tautol-
ogical bog (Pomerance, 1982, p. 2). But the definitional

problem will remain, for the term 'people' is included in many international legal instruments.

Is it enough to regard 'people' as belonging only to the total population of a state? Is it possible to define 'people' freed from the state-territorial limitation? Might there not be wisdom in simply having all sub-state groups define themselves as they themselves desire, declaring their 'peoplehood' if such fits with their sense of group 'self'? As Cobban (1945, p. 48) put it: 'Any territorial community, the members of which are conscious of themselves as members of a community, and wish to maintain the identity of their community, is a nation'. At issue here is whether or not the 'territorial community' is the total population of a state, bound together by a nationalism or some similar sentiment, or a sub-state population that, in turn, might be regarded as merely a regionalism by those who are linked to the totality of the state (Knight, 1982).

Should such sub-state groups simply be regarded as minorities? Does it make a difference if a 'people' are a 'people' versus a 'minority'? Capotorti (1977) defined a minority as:

> A group numerically inferior to the rest of the population of a State, in a non-dominant position, whose members - being nationals of the State - possess ethnic, religious or linguistic characteristics differing from those of the rest of the population and show, if only implicitly, a sense of solidarity, directed towards preserving their culture, traditions, religions or language.

We can differentiate between 'well-established historical minorities' (who may not have been minorities in the past) and 'immigrants' and observe that various rights of minorities within states are laid out in several legal instruments (see, e.g. Henkin, 1981). But there remains the issue of to whom self-determination can be applied. The issue becomes even more complicated when linked with the debate over individual versus group rights. Richard Falk (cited in Davies, 1985, p. 776) concluded that:

> the original impulse of the doctrine of self-determination, what led to its formation as a principle, has to do with a basic affirmation of respect for the inherent dignity of individuals and groups, and that fundamentally the internal application of the doctrine of self-determination is a human rights claim ... .

Individuals <u>and</u> groups? Certainly, the Covenant on Civil and Political Rights and the Covenant on Economic,

Social and Cultural Rights both declared self-determination to be a right; however this generally has been taken, at least in the West, to apply to persons belonging to minorities as individuals but not to minorities as collectivities. Of note, the Covenants did not confine self-determination only to colonial peoples. Indeed, they declared that all states should 'promote the realization of the right of self-determination'. This can be taken to refer to peoples in colonized or otherwise dependent territories and also to 'people' within sovereign states.

Support comes for the latter proviso from the non-binding but persuasive 1975 Helsinki Final Act, which included an outwardly clear statement on self-determination (Knight and Davies, 1987, document 31). If the signatories intended their statements to apply only to external self-determination, why was it included and so emphatically stated? From a legal perspective, Cassese (1981, p. 94) felt that while, 'to a very great extent, the principle of external and internal self-determination has already been realized in Europe (as well as in the United States and Canada) and, therefore, it would be pointless to codify and reaffirm it in some sort of regional European instrument', yet he noted that situations still exist in Europe 'which come within the purview of a broader concept of self-determination'. This conclusion might be interpreted to mean that self-determination can apply to sub-state minorities, but there was no clear indication in the Helsinki Final Act that the right can definitely be extended to minority groups within states. This point surely will continue to be debated. A critical issue not grappled with in the Act and other legal documents pertains to those 'people' who never voluntarily gave up their original status and who do not, or not completely, identify themselves with the state within which they live. These people generally are referred to as indigenous peoples.

Peoples, Not Minorities

Representatives of indigenous peoples from many parts of the world, acting in concert, declared (in the words of a 1981 draft declaration) that:

> In general, all indigenous peoples are entitled to self-determination and to recognition as nations. ... Therefore, indigenous peoples ought not be considered as minorities or social classes ... (Knight and Davies, 1987, document 482),

Further, indigenous peoples have objected to the International Labour Organization's Convention 107 (of 1957) for its reference to mere 'tribal populations', use of paternalistic language, and emphasis on integration and assimilation.

Some 'Europeans' ask how indigenous peoples can 'prove' they are 'people' or 'nations'? Such a stance must be challenged, for why should indigenous peoples 'prove' such to anybody, other than perhaps ourselves? The more important question is, how can 'Europeans' claim indigenous peoples are not 'peoples' or 'nations'? By what right do they exclude indigenous peoples from being so defined? In approaching the issue a 1981 report to the U.N. Human Rights Commission is useful. The report defined a people as 'a social entity possessing a clear identity with its own characteristics', including some 'relationship with a territory, even if the people in question has been wrongfully expelled from it and artificially replaced by another population' (Cristescu, 1981). A problem with this definition is that certain 'ethnic' minorities may also meet the cultural and territorial criteria suggested by Cristescu, despite his acknowledgement that 'a people should not be confused with ethnic, religious or linguistic minorities, whose existence and rights are recognized in Article 27 of the International Covenant on Civil and Political Rights'. But, to repeat a point made earlier, the right of persons belonging to minorities as individuals has been recognized, yet it is clear that indigenous peoples desire to exercise their right to self-determination as groups, as peoples, not as individuals.

A report to the U.N. Working Group on Indigenous Populations in 1983 concluded that:

> Indigenous populations, communities, peoples and nations are those which, having a historical continuity with pre-invasion and pre-colonial societies that developed on their territories, consider themselves distinct from other sections of the societies now prevailing on those territories or parts of them. They form at present non-dominant sectors of society and are determined to preserve, further develop and transmit to future generations their ancestral territories, and their ethnic identity as the basis of their continued existence as peoples, in accordance with their own cultural patterns, social institutions and legal systems (United Nations, 1983).

The use of the word 'ethnic' in this statement represents an outsider's view, because indigenous peoples are insistent that they are neither 'ethnics' nor 'minorities' but 'peoples' and, as such, are to be regarded as 'nations'.

Thus, for instance, the Dene of Canada's Northwest Territories have declared that:

> We the Dene ... insist on the right to be regarded by ourselves and the world as a nation. ... . Our plea to the world is to help us in our struggle to find a place in the world community where we can exercise our right to self-determination as a distinct people and as a nation (Dene Declaration, in Watkins, 1977, pp. 3-4).

There is an historical foundation for indigenous peoples calling themselves peoples and nations.

Some historical considerations

The roots of indigenous peoples' rights in international law date to the sixteenth century when Francisco de Vitoria recognized that the Indians of the Americas had inherent rights to life and property, by reason of their humanity. Vitoria's treatise (Nys, 1917, p. 128) concluded that:

> The aborigines undoubtedly had true dominion in both public and private matters, just like Christians ... . The aborigines in question were true owners, before the Spanish came among them, both from the public and private point of view.

Vitoria's outlining of the basic concepts of the rights of indigenous peoples has 'survived the rigours of time practically intact' (Davies, 1985a, p. 20). Vitoria objected to European powers claiming new territories by right of discovery, for, he stated, the latter proviso pertained to 'that what belongs to nobody is granted to the first occupant' whereas there can be no such claim when it involved the 'seizure of the aborigines any more than if it had been they who had discovered us'.

Vitoria's writings were persuasive and influenced, among others, Pope Paul in 1537 and, a century later, Hugo Grotius (Cohen, 1942). Despite being a concern of international law for several centuries, the exact nature of the legal status of indigenous peoples in their dealings with the societies that dominate them remains unresolved. Davies (1985a, pp. 24-43) suggested various elements of international law that need to be explored: sovereignty, treaties, guardianship, and territorial rights.

Did or do indigenous peoples have sovereignty? The question is simple, the answer difficult. There is a rich series of legal pronouncements on this question, with the 1831 decision of the U.S. Supreme Court being essential. In

The Cherokee Nations vs. The State of Georgia the court decided that 'The Cherokees are a state' and that the actions of the U.S. Government 'plainly recognize the Cherokee nation as a state'. And, in 1832, in the judgement handed down in the case Worcester vs. Georgia, the Court ruled that 'The very term "nation", so generally applied to them, means "a people distinct from others"' and indicates that the U.S. 'has adopted and sanctioned the previous treaties with the Indian nations, and consequently, admits their rank among those powers who are capable of making treaties'.

Despite these rulings and many others, which have been incorporated into international law, there was no general acknowledgement among the colonial and successor states as to whether or not agreements with indigenous peoples were treaties in the classical international legal sense. Many treaties were signed in several states, and indigenous peoples have consistently argued that their forefathers, as sovereign nations, made treaties with sovereign nations. Such a stance finds support in the U.S. Supreme Court ruling on Worcester vs. Georgia:

> The words "treaty" and "nation" are words of our own language, selected in our diplomatic and legislative proceedings, by ourselves ... We have applied them to Indians, as we have applied them to the other nations of the earth. They are applied to all in the same sense.

Treaties or not, some indigenous peoples and their territories were 'conquered' by armed might, with people being massacred or suffering from forced relocations; others had initial friendly contacts, but these often led to decimation from introduced diseases. All too often, the immigrant Europeans held the attitude that indigenous peoples were not 'civilized' (one which unfortunately prevails still) and so could simply be set aside, with force if necessary, and their lands declared terra nullius and thus 'acceptable' for new settlement. The end result of European migrations was the indigenous peoples' loss of control over their territories. New political identities were created according to the limits of the new political territories, identities often not accepted by the indigenous peoples.

The Centrality of Land

Indigenous people have argued that they could never have given up their rights to their lands, to their territories, for the latter were not theirs to give up. In the words of a Chippewa chief, in prayer:

> Great Grandfather, you gave us the land and its resources, you made us one with the birds, the animal life, the fish life; you made us one with nature itself. This is our aboriginal right. It is a right that no government can interpret for us. Because you gave it to us, no man has a right to take it away from us (Plain, 1985, p. 40).

Land is at the base of disputes between indigenous peoples and the particular settler society that now dominates their lives and their territories. Europeans see land as a commodity; indigenous peoples see land as central within the circle of life. Thus a chief within the Iroquois Nation, to the question, what are aboriginal rights, concluded that:

> They are the law of the Creator. That is why we are here; he put us in this land. He did not put the white people here; he put us here with our families, and by that I mean the bears, the deer, and the other animals. We are the aboriginal people and we have the right to look after all life on this earth. We share land in common, not only among ourselves but with the animals and everything that lives in our land. It is our responsibility ... (Lyons, 1985, p. 19).

The linkage between identity and territory is nicely summed up by the Maori term turangawaewae which, literally, means standing place for the feet, and implies 'the rights of a tribal group in land and the consequential rights of individual members of the group' (Kawhara, 1979, p. 3).

Identity, territory, rights, responsibilities - four linked fundamental concepts. The word control is missing from the list: control of the land (including renewable and non-renewable resources contained therein) and also spiritual, social, economic and political control. Without control, the peoples' rights and responsibilities cannot be fulfilled and identity within territory is threatened. The word control implies self-determination. But indigenous peoples for the most part lack full control over their lives and thus they exist essentially as dissenting voices within states. As such, indigenous peoples face the full range of means available to states for 'dealing' with them, from liberal encouragement to tyrannical threats of violence.

Asserting 'Sovereignty'

The earlier discussion of territorial acquisition by Europeans and their assertion of sovereignty indicates that indigenous peoples today are generally viewed by Europeans as lacking sovereignty. Something is missing from the textbook criteria for the attainment of sovereignty, however, and that is the conception of sovereignty as held by some indigenous people. It was nicely enunciated by a North American Indian who recognized that the exercise of sovereign powers by Indian governments had been restricted to a considerable extent, but he felt that the United States and other states had recognized 'the inherent sovereignty of Indian nations and their right to self-government'. Kicking-bird (1984, p. 48) continued:

> We know that Indian governments are sovereign because 1. Indians feel they are sovereign. 2. Indian governments historically exercised sovereign powers. 3. Other nations have recognized the sovereignty of Indian governments.

Are such statements merely reflective of romanticism? Should not such claims simply be dismissed for they have no foundation in European or European-derived (i.e. non-indigenous) law? But the issue of international recognition and assertion of 'sovereignty' are important. Tied to these issues has been the recent realization of significant linkages among and between indigenous peoples from many parts of the world via inter-group coalitions. One such coalition is the World Council of Indigenous Peoples which was established in 1974, with membership today extending to 31 states in five regions of the world: North America, Central and South America, the Pacific (including New Zealand and Australia), Northern Europe and Asia. In 1981, in Melbourne, Australia, the Council drafted an International Covenant on the Rights of Indigenous Peoples. Its language is instructive:

> Article 1. All peoples have the right to self-determination. By virtue of that right Indigenous Peoples may freely determine their political status and freely pursue their economic, social and cultural development.
> Article 2. The term Indigenous People refers to a people (a) who live in a territory before the entry of a colonizing population, which colonizing population has created as new state or states or extended the jurisdiction of an existing state or states to include the terri-

tory, and (b) who continue to live as a people in the territory and who do not control the national government of the state or states within which they live.

Article 3.   One manner in which the right of self-determination can be realized is by the free determination of an Indigenous People to associate their territory and institutions with one or more states in a manner involving free association, regional autonomy, home rule or associate statehood as self-governing units. Indigenous People may freely determine to enter into such relationships and to alter those relationships after they have been established.

Article 4.   Each state within which an Indigenous People lives shall recognize the population, territory and institutions of the Indigenous People ... .

These are grand ideas, boldly stated, but most states will be quite reluctant to accept most of what is called for.
    In 1982 the U.N. Commission on Human Rights created the Working Group on Indigenous Populations – the word 'peoples' was too suggestive for the U.N. to use.   Many delegations of indigenous peoples attend the annual meetings of the Working Group, along with observers representing governments, several UN specialized agencies, and those indigenous organizations that hold UN consultative status (including The World Council for Indigenous Peoples, The Indian Law Resource Center, and the International Indian Treaty Council).   In 1985 over 200 indigenous representatives from around the world participated in the meeting. The Working Group of Indigenous Populations has been especially important since it has become the principal internationally recognized forum where indigenous peoples may speak about their grievances directly.   In assessing the importance of the Working Group, Davies (1986) concluded that 'the particular vulnerability of indigenous peoples has been recognized by virtually all governments and the working group has begun the important process of standard setting to provide certain basic protections'.   Of concern, however, is that the existence of this vitally important body to indigenous peoples has recently been threatened by the severe financial problems faced by the U.N. (United Nations, 1986).
    International efforts to gain recognition represent one important way by which self-determination can be acknowledged for indigenous peoples.   For some groups this represents perhaps the only manner by which their case can

be heard internationally. The repression some indigenous peoples face is considerable, as in Guatemala. However, recognition and participation at the international level has meant that representatives of such indigenous peoples have been able to reveal the severe human rights violations caused by the state governments that dominate their lives. Some of the spokespersons have risked their lives in order to reach Geneva. There are difficulties even for people in 'open' societies in getting to Geneva, however, since governments generally provide the travel costs for 'approved' representatives of their states' indigenous peoples but dissident groups have been prevented from attending the meetings due to lack of funds. Nevertheless, the fact of international involvement clearly represents a major break-through for indigenous peoples. In a sense, the recognition is seen to be an assertion of 'sovereignty', even if the latter is more a matter of the mind and spirit than political control.

If it is accepted that the right of self-determination can be applied to indigenous peoples then there are immediate problems in coming to terms with just what is being called for. For instance, indigenous peoples in Canada indicate they want 'self-determination' yet there is no agreement on what that means. Some want local social/political/economic control, control that clearly would be within provincial territorial bounds, that is, essentially local government. It should be noted, however, that many groups now view such controls with great suspicion following the James Bay agreement whereby native peoples in northern Quebec ceased being Federal Government responsibilities as they lost any claim to aboriginal title and became simply citizens of Quebec province in return for being granted local government, selected land and hunting rights, and a financial settlement. Other indigenous people call for self-determination whereby politico-territorial controls would be set apart from the provinces as a new parallel level of government, but still within sovereign Canada. Most provincial leaders currently frown on this proposal, however, for the granting of such concessions may involve giving up further territory - by creating enclaves within provincial boundaries - and certain controls that they see as 'theirs', even though the territories in question now mainly exist as 'reserves' under the Federal Government's jurisdiction. Still other indigenous people want some new form of self-determination that would at once be within the State of Canada but with the right to establish and maintain international linkages. The latter conception would challenge the Canadian State's sovereign rights and, possibly, under some proposals, territorial integrity although this does not necessarily mean secession. Finally, there is a (distinct) minority who would like to see the total attainment of sovereignty for certain territories, but success for such a goal is a most unlikely proposition. The debate continues.

Indigenous peoples face many hurdles, some of which are noted here. In Canada, as elsewhere, including Scandinavia, New Zealand, and the U.S.A., there is a fundamental question of just who is an indigenous person. In Canada, for instance, there are 'status' and 'non-status' Indians, Metis (descendants of inter-racial marriages who have developed a collective identity as a separate people), and Inuit, each with differing rights, for some signed treaties whereas others did not. Identity is gained by birth, although in the instance of New Zealand's Maoris, they may choose to register to vote either on a special Maori electoral roll or on the general roll. There also is the issue of urbanized indigenous peoples, many of whom are quite alienated from the dominant societies. Some have a desire for being rooted in their people's land once more, even if they have not been there during their lifetime (Manuel, 1973; Ihimaera 1972) and may be prevented from 'returning' by government decrees. Problems of cohesion exist because traditional or special territories may be quite dispersed and fragmented. Land 'claims' by indigenous peoples are being made in many states and are having varying degrees of success (Raby , 1974). In some states there is the issue of regional dominance. For example, the Norwegian interior region where the Sami are numerically dominant is not where the majority of Sami now live. And the Sami represent another problem, that of a people who are divided by several states' international boundaries.

The issues are many. In approaching them two explanatory models can be noted. First, there is a class model which identifies indigenous peoples as constituting the most disadvantaged group within society. Recognition of the low socio-economic position can lead to the formulation of assimilationist "catch-up" policies that are designed to integrate the indigenous peoples into the dominant society, with a certain number becoming involved in governmental bureaucracies and the development of new elites who will deal with those bureaucracies. Second, there is a centre-periphery model which recognizes that indigenous peoples are generally located in peripheral underdeveloped regions that are poor, by every measure, due in part to central government inaction and neglect. It is particularly out of this second mode of thinking that self-determination claims are being generated, for indigenous peoples in many states desire a transfer of power from the centre to themselves so that they can have control over themselves in the lands they hold or claim.

Conclusion

Indigenous peoples have suffered profound attacks on their land and on their distinctive ways of life. By remarkable doggedness, most indigenous peoples have been able to fight cultural and even physical extinction by utilizing various economic strategies (Stea and Wisner, 1984). As numerical and spiritual strength and confidence have been regained, after decades and centuries of dispossessions and ill-treatment, international linkages and international actions are combining to give a new sense of purpose to many indigenous peoples. Throughout the world indigenous peoples have started 'organizing to survive' (Clay, 1984). They realize they are alone no longer and thus their plight now is not simply an intra-state matter. International norms for human rights' observances have been enunciated and are being used by indigenous peoples as they seek to better their own lot - and that of those indigenous peoples who cannot speak internationally for themselves.

Despite differences there are commonalities. George Manuel's phrase the 'Fourth World' seeks to capture the commonalities, in that he felt all indigenous peoples were bound by a shared concern for caring human-land relationships and by the lack of control over their lives (Manuel, with Posluns, 1974). Increasingly this latter issue is being addressed. The remarkable persistence of indigenous peoples has resulted, among other things, from strength gained from age-old spiritual beliefs. Contemporary calls for self-determination for and by indigenous peoples arise from an anxiety over their future in lands that often are threatened by European intrusion but also, to a large degree, from cultural and spiritual concerns. As such, the calls by indigenous peoples for self-determination will now not readily be ignored. A political geographic analysis of their calls for self-determination and the resulting patterns and processes remains to be accomplished.

Acknowledgements

This paper is based on research for a project on self-determination that has been supported by the Social Sciences and Humanities Research Council of Canada. The Council's support is gratefully acknowledged. The author also is grateful to Professor Maureen Davies and Mr. Brian Gottheil for their reactions to a draft of the paper, and to Mrs. Hazel Anderson for typing drafts of the paper under tight deadlines.

References

Brierly, J.L. (1963) The Law of Nations, 6th ed, Oxford
    University Press, Oxford
Capotorti, Francesco (1977) Study on the Rights of Persons
    Belonging to Ethnic, Religious and Linguistic Minori-
    ties, U.N. Docs. E/CN.4/Sub.2/384 and Add. 1-7
Cassese, Antonio (1981) 'The Self-Determination of peo-
    ples', in L. Henkin (ed.), The International Bill of
    Rights, Columbia University Press, New York, 92-113,
    416-427
Clay, Jason W. (1984) 'Organizing to Survive', Cultural
    Survival Quarterly, 8 (4), 2-5
Cobban, A. (1945) National Self-Determination, Oxford
    University Press, Oxford
Cohen, F. (1942) 'The Spanish Origins of Indian Rights in
    the Law of the United States', Georgetown Law Journal,
    31, 1-4
Crawford, J. (1979) The Creation of States in International
    Law, Clarendon Press, Oxford
Cristescu, Aureliu (1981) Report to the U.N. Human Rights
    Sub-Commission, U.N. Doc. E/CN.4/Sub.2/404
Davies, Maureen (1985a) 'Aspects of Aboriginal Rights in
    International Law', in Bradford, W. Morse (ed.),
    Aboriginal Peoples and the Law. Carleton University
    Press, Ottawa, 1-47
Davies, Maureen (1985b) 'Aboriginal Rights in International
    Law: Human Rights', in Bradford W. Morse (ed.),
    Aboriginal Peoples and the Law, Carleton University
    Press, Ottawa: 745-794
Davies, Maureen (1986) 'The U.N. Arrears Crisis: Its
    Implications for Human Rights', Human Rights Internet
    Reporter, 11, (2), pp. 6-7
Henkin, L., ed. (1981) The International Bill of Rights,
    Columbia University Press, New York
Ihimaera, Witi (1972) Pounamu Pounamu. Heinemann, Auckland
Indian Law Resource Center (1984) Indian Rights - Human
    Rights: A Handbook for Indians on International
    Human Rights Complaint Procedures, Indian Law Resource
    Centre, Washington, D.C.
Kawhara, I.H. (1979) Land as Turangawaewae: Ngati Whatua's
    Destiny at Orakei. New Zealand Planning Council,
    Wellington: Planning Paper No. 2
Kickingbird, Kirke (1984) 'Indian Sovereignty: The
    American Experience', in Leroy Little Bear, Menno
    Boldt, and J. Anthony Long (eds.), Pathways to
    Self-Determination - Canadian Indians and the Cana-
    dian State, University of Toronto Press, Toronto,
    46-53.
Knight, David B. (1982) 'Identity and Territory: Geo-
    graphical Perspectives on Nationalism and Region-

alism', Annals of the Association of American Geographers, 71, 514-532

Knight, David B. (1984) 'Geographical Perspectives on Self-Determination', in Peter Taylor and John House (eds), Political Geography: Recent Advances and Future Directions, Croom Helm, London; Barnes and Noble, Totowa, N.J., 168-90

Knight, David B. (1985a) '"Minorities" and Self-Determination', in David B. Knight (ed.), Our Geographic Mosaic, Carleton University Press, Ottawa, 139-47

Knight, David B. (1985b) 'Territory and People or People and Territory: Thoughts on Post-Colonial Self-Determination', International Political Science Review, 6, 248-72

Knight, David B. and Davies, Maureen (1987) Self-Determination: An Interdisciplinary Annotated Bibliography, Garland Publishing, New York

Lyons, Oren (1985) 'Traditional Native Philosophies Relating to Aboriginal Rights', in Menno Boldt and J. Anthony Long (eds), The Quest for Justice: Aboriginal Peoples and Aboriginal Rights, University of Toronto Press, Toronto

Manuel, George (1973) personal communication

Manuel, George with Posluns, Michael (1974) The Fourth World: An Indian Reality. Collier-Macmillan, Toronto; Macmillan, New York, 19-23

Nys, E. (1917) Relictiones: De Indis et De Jure Belli, Carnegie Classics of International Law, Washington, D.C.

Plain, Fred (1985) 'A Treatise on the Rights of the Aboriginal Peoples of the Continent of North America', in Menno Boldt and J. Anthony Long (eds.), The Quest for Justice: Aboriginal Peoples and Aboriginal Rights, University of Toronto Press, Toronto, 31-40

Pomerance, Michla (1982) Self-Determination in Law and Practice: The New Doctrine in the United Nations, Martinus Nijhoff, The Hague

Raby, S. (1974) 'Aboriginal Territorial Aspirations in Political Geography', Proceedings of the International Geographical Union Regional Conference, New Zealand Geographical Society, Palmerston North, 169-74

Shafer, Boyd C. (1972) Faces of Nationalism, Harcourt Brace Jovanovich, New York.

Stea, David and Wisner, Ben (1984) 'Introduction' to special issue on 'The Fourth World: A Geography of Indigenous Peoples', Antipode, 16 (2), 3-12

United Nations (1954) Report on the Tenth Session of the Commission on Human Rights, 18 ESCOR Supp. 7, U.N. Doc. E/2573

United Nations (1960) <u>Declaration on the Granting of Independence to Colonial Countries and Peoples</u>, U.N. Resolution 1514 (XV)

United Nations (1983) <u>Ideas for the Definition of Indigenous Population, from the International Point of View</u>, Commission on Human Rights, Sub-Commission on Prevention of Discrimination and Protection of Minorities, Working Group on Indigenous Populations, U.N. Doc. E/CN.4/Sub.1/AC.4/1983/CRP.2

United Nations (1986) <u>Report of the Secretary-General on the Current Financial Crisis of the U.N.</u>, U.N.A./40/1102

Watkins, Mel (1977) <u>Dene Nation: The Colony Within</u>, University of Toronto Press, Toronto

Chapter 9

EVOLVING REGIONALISM IN LINGUISTICALLY DIVIDED BELGIUM

Alexander B. Murphy

In 1974 the American political scientist Aristide Zolberg
published an article entitled 'The Making of Flemings and
Walloons' which traced the development of ethnolinguistic
identities in Belgium from the emergence of the Belgian
state in 1830 to the beginning of World War I (Zolberg,
1974). Zolberg pointed to the lack of organized, linguistically based group sentiment in Belgium in 1830, and then
sought to explain the growth of Walloon and Fleming identities against the backdrop of socio-economic and political
circumstances in the nineteenth and early twentieth centuries. A parallel development which accompanied this
transformation might be described as 'the making of Flanders
and Wallonia'. This involved the creation of linguistically
defined regions in Belgium of conceptual, and later functional and administrative, significance. Far from being
merely the result of ethnolinguistic tensions, the formation
of these regions has been inextricably tied to the development of linguistically based group consciousness in Belgium.
In fact, the shift to the use of linguistic criteria as a
basis for subdividing Belgium has been both a response to
growing ethnolinguistic consciousness and an agent in
structuring intergroup and intragroup relations.

The development of a linguistically based regional
conception of Belgium is symbolized by the history of the
terms 'Wallonia' and 'Flanders'. At the time of the Belgian
Revolution in 1830, there were no generally accepted
toponyms for the language regions of Belgium. The only
regional divisions shown on administrative maps from that
period were the nine provinces, direct progeny of the
departments into which Belgium had been divided under French
rule (Luykx, 1959, plate 13). The term 'Wallonia' had not
yet been coined (Henry, 1974), and 'Flanders' was used
either to refer to the territory of the medieval County of
Flanders or to the Belgian provinces of East and West
Flanders (Gysseling, 1975). In the ensuing 150 years,
'Wallonia' came to signify the largely French-speaking part

of Belgium south of the Romance-Germanic language line which bisects the country, while 'Flanders' came to be used to refer to the predominantly Dutch-speaking area north of the linguistic boundary. Flanders and Wallonia are now the primary administrative regions in a state which during the past twenty-five years has been formally partitioned along linguistic lines (see Figure 9.1). Together with the bilingual communes of Brussels and a small German-speaking area in the east, Flanders and Wallonia are constitutionally recognized linguistic regions with substantial autonomy in a wide range of cultural, social, and economic matters. Although the provinces remain as administrative units, the linguistic regions have assumed primary importance in Belgium's internal administrative structure.

In short, during Belgium's relatively brief history as an independent country, a transformation has occurred in the conceptual and formal regionalization of the state. From a beginning in 1830 as a highly centralized state in which territorial identities were either local, or in the case of a political and social elite, national (in the sense of state nationalism), Belgium has seen the rise of strong regional identities based on linguistic criteria and the institutionalization of those identities in formal territorial units. The units play a visible and significant role in social, political, and economic relations within the state.

Regional Conceptions and Territoriality

Although the historical record confirms that linguistic distributions were not always an important basis for the conceptual partitioning of Belgium, the significance of language regions in modern Belgium has tempted many to assume that Wallonia and Flanders are valid units for analyzing ethnolinguistic relations throughout Belgium's history as a sovereign state. The problem is exemplified by the following quotation which introduces a recent article on Belgium in the Financial Times (Dickson, 1986):

> Asked recently why Flemish-speaking Flanders and French-speaking Wallonia should ever have united to form a single sovereign state, a Belgian commentator noted unequivocally, "It was an historical accident".

The question as presented, however, cannot be answered because it assumes a Swiss-style voluntary banding together of distinct ethnolinguistic regions which simply did not exist at the time of Belgian independence. Such an assumption ignores the fact that the Belgian case is fundamentally different from the Swiss situation. In Belgium, language

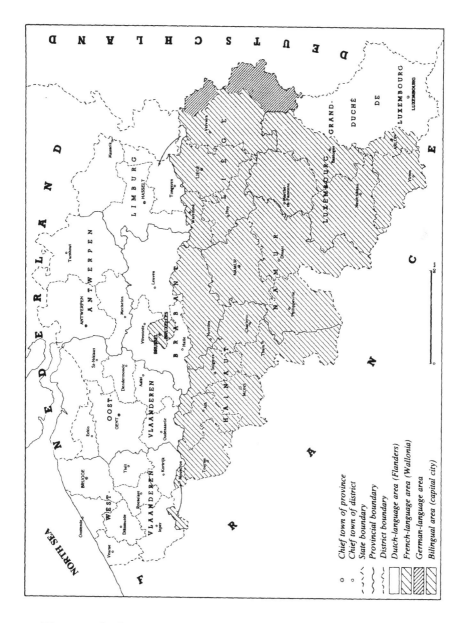

Figure 9.1. The linguistic and administrative
divisions of Belgium. Source: derived from
Belgian Information and Documentation
Institute, 1985. <u>Atlas of Belgium</u>. Brussels:

areas did not develop into significant, well-defined political/cultural regions until well after the formation of the state, with all that that implies in terms of territorial conflict.   In Switzerland, by contrast, language regions already had a reasonable degree of autonomy at the time of confederation because of the relatively close areal correspondence between zones of language change and the boundaries of certain cantons (McRae, 1983).   Exceptions, such as the Jura canton, where there has been significant tension, serve only to strengthen the argument.

If the above-quoted misconception about regionalization were limited to the popular press, we could perhaps dismiss it as one more geographical lacuna in the knowledge of the general population.   Unfortunately, however, the scholarly literature on Belgium is also replete with explicit or implicit references to Wallonia and Flanders as analytical constants which have not changed in meaning or significance throughout Belgium's history.   What is forgotten is that there is nothing inevitable about regions. Rather, regions are human constructs which come about as the result of the conceptual division of the world on the basis of a set of chosen criteria   (Knight, 1982, p. 517). Moreover, the selection of criteria inevitably changes with the geographical, social, and historical setting.   Regionalization (and I am using the term here to refer to both conceptual and formal processes of dividing up the world) can follow physiographic, economic, political, or cultural lines, but need not follow any one or the other, and almost certainly will not remain constant over time as the significance and distribution of the underlying phenomena change. At the same time, the ways in which people conceptually and formally partition the world are of fundamental importance in human relations.   This is particularly true with respect to ethnic groups with autonomist or nationalist ambitions because of the important ideological and functional roles which territory and region can play for such groups.

While some ethnic groups lack a well defined territorial base and accompanying regional identity, most of the significant ethnic tensions in the modern world arise between groups associated, by themselves or by others, with distinctive regions within states.   As such, the concepts of region, space, and territory are of fundamental importance for interethnic relations.   In fact, ethnic groups are frequently defined in regional terms because territorial considerations are so fundamental to the maintenance of group boundaries and, by extension, to ethnic group survival (Alverson, 1979, p. 13).   Moreover, it is the lack of areal correspondence between ethnic patterns and political patterns which is at the heart of most interethnic tension (Mikesell, 1983).   Finally, relative situational context is an important variable in ethnic relations, particularly as it relates to patterns of economic development and differentiation.

Given this strong spatial and territorial component to ethnicity and ethnic relations, the subject is one to which geographers have much to contribute. Unfortunately, until quite recently few geographers have been sufficiently concerned with the causes or processes of ethnic tensions to move beyond the descriptions of ethnic patterns and traits which form a part of most regional geographies or the cursory examination of ethnic issues which appears in many political geography textbooks (Knight, 1982). There are, however, a few notable exceptions. During World War I, Leon Dominion published a major work on the distributions of linguistic and national groups in Europe (Dominion, 1917), and Isiah Bowman, among others, brought a geographer's knowledge of these matters to the Paris Peace Conference. Considerably later, a few geographers began to focus in detail on the causes and consequences of ethnic heterogeneity within particular politically organized areas (e.g. the study of Malaya by Ginsburg and Roberts, 1958).

More recently, a number of geographers have turned their attention to these matters, as suggested by the range of publications of the contributors to this book. Moreover, a number of students of ethnicity in other disciplines have brought spatial concepts to bear in seeking to explain the rise of ethnic tensions. In fact, two of the most widely discussed recent theories of ethnic conflict, Hechter's internal colonialism (Hechter, 1975) and Nairn's uneven development (Nairn, 1977), are both concerned, at least indirectly, with spatial differentiation as a causal factor in the development of tensions.

Spatially derived studies of this sort are concerned with the inter-relationships between cultural, social, economic, and political patterns, and the implications of those relationships for the development of ethnic tensions. Their contribution has been to identify a number of the key elements associated with the rise of ethnic conflict. The uneven development and internal colonialism theories have left themselves open to criticism, however, because they fail to provide much information about the timing, location, or intensity of ethnic insurgences (Orridge and Williams, 1982, p. 35; Smith, 1982, pp. 20-24). What is missing, in part, is an understanding of when and how spatial units themselves acquire meaning and significance and an appreciation of the implications of that process. Focusing on these questions is vital in order to avoid the pitfalls of explaining historical developments through present compartmentalizations of space. Thus, although economic differentiation is an important explanatory element in the rise of ethnolinguistic tensions in Belgium during the nineteenth century, basing an analysis solely on the idea of an industrialized, economically strong Wallonia dominating an agrarian, backward Flanders ignores the true complexities of the situation. In fact, throughout much of the nineteenth century, the present regions of Wallonia and Flanders did not constitute conceptual or economic entities in any meaningful sense. For example, the Flemish province of Limburg and the Walloon province of Luxembourg occupied a rather similar economic position within the Belgian state, as suggested by the fact that they were the only two provinces in Belgium with more than 60 per cent of the workforce engaged in agriculture at the time of the 1846 census (Annuaire Statistique, 1870,pp. 38-39).

To move beyond these problems it is necessary to understand how Flanders and Wallonia came to acquire such a degree of conceptual and formal significance that Luxembourg's ethnolinguistic ties to the rest of Wallonia became more important than its socio-economic similarity to Limburg. In effect, what is needed is a more explicit concern for the ways in which people and institutions develop loyalty to and assert control over particular territories. Robert Sack (1981, 1983) uses the term territoriality to refer to these processes. For Sack, territoriality involves 'the attempt to affect, influence, or control actions and interactions (of people, things, and relationships) by asserting and attempting to enforce control over a geographic area' (Sack, 1983, p. 55). In his studies of territoriality, Sack demonstrates that the preoccupation of many geographers with issues of spatial distribution and contact has meant that little attention has been devoted to human efforts to control space. Yet it is precisely this process which is behind the transformation of linguistic areas in Belgium into regions of social and political significance.

An Overview of the Belgian Case

There is an extensive body of literature documenting the historical development of ethnolinguistic consciousness and language group relations in Belgium (see the bibliographies of Verdoodt, 1977, 1983). Rather than attempting to summarize that literature in all its complex detail, the concern here is to highlight the most important developments in the process by which language areas acquired conceptual and formal significance in an effort to demonstrate the importance of that process for evolving intergroup relations in Belgium. As noted previously, at the time of Belgian independence there was no significant regional consciousness based on linguistic criteria. This is not to say that language was an unimportant issue in the Belgium of 1830. In fact, despite a constitutional guarantee of freedom of language use (article 23), the Belgian Revolution was primarily a Francophone affair, and the Francophones held the upper hand in the newly created state (Polasky, 1981). Nevertheless, this did not translate into regional consciousness based on language, in part because a significant percentage of the Belgian revolutionaries came from a Flemish bourgeousie which had adopted French as its language by the time of Belgian independence, and in part because the seeds of a Belgian-wide national will had already been sown by more than two hundred years of shared history (Stengers, 1981; De Schryver, 1981).

Belgium emerged as a highly centralized unitary state with a French-speaking upper class. French became the language of government, business, and higher education. The dominant economic region was situated in part of the French language area. Over the course of the next half century, a Flemish movement gradually took form in reaction to the exclusion of the Dutch language from the upper levels of the Belgian government, economy and society (Fredericq, 1906). From its beginning as a largely literary movement led by Flemish intellectuals, it became more and more political during the latter part of the nineteenth century. During this period, demands centred on official recognition of the Dutch language and on the right of Dutch speakers to use their language in the army, the courts, and the government (Clough, 1968, pp. 130-174). These demands led to the enactment of the first important linguistic laws during the last quarter of the century (Maroy, 1966, pp. 457-460).

The position of Dutch in Belgium was the central issue of the early Flemish movement. Not surprisingly, the Dutch language area began to take on some degree of conceptual significance during this period. In fact, during the course of the second half of the nineteenth century, the term Flanders came to be used to refer to all of Dutch-speaking Belgium. The language laws which were passed were, by their terms, territorially limited in application to the northern part of Belgium. Moreover, the growing success of the leaders of the Flemish movement in forging a degree of Flemish regional identity led, by the end of the nineteenth century, to the beginnings of attempts by a few Walloon civil servants and intellectuals to encourage regional consciousness in the development of the South (Schreurs, 1960, pp. 10-15).

Despite these developments, by the turn of the century regional consciousness was still quite limited in Flanders and Walloon regional identity was virtually nonexistent. The subject catalogue at Belgium's national library does not contain any listings under the headings 'Wallonie', 'Wallon', or 'Wallons' of works appearing before 1900 with a specific reference to the region of Wallonia in the title. Only a handful of pre-1900 works appear under the headings

'Vlaanderen', 'Vlaamse', or 'Vlamingen' which mention Flanders in the title in reference to the Dutch-speaking part of Belgium. In fact, before the turn of the century virtually no one had thought of Flanders and Wallonia as functional or formal units, and even the leaders of the Flemish movement espoused loyalty to the Belgian state (De Schryver, 1981, p. 27). Although the language laws of the time applied only to Flanders, they were not significant expressions of Flemish territoriality because they did not serve to exclude French in Flanders. Rather, they were aimed at giving Flemings the right to use their mother tongue in certain situations (Maroy, 1966, pp. 457-460). Their territorial limitation was tied to the physical distribution of Flemings and the political infeasibility of enacting legislation protecting the rights of Dutch speakers in areas which were overwhelmingly French in speech.

The period just before and during World War I saw a significant change in strategy by the leaders of the Flemish movement (and even to some degree by their counterparts in the nascent Walloon movement). The focus shifted from the rights of individual language speakers to the collective rights of the inhabitants of the language regions. The Flemish activists were concerned about the frequent disregard of the language laws which had been passed, the persisting disadvantaged position of Dutch speakers in Belgium, and the growing Frenchification of the Brussels population. Even though Dutch speakers were in a numerical majority in the country, suffrage laws of the time effectively put them in a minority position. A territorial approach provided a possible way of gaining political leverage. Also, a territorial strategy could provide a basis for promoting Flemish sentiments throughout the northern provinces among a population divided by dialectical and socio-economic differences. The leaders of the Flemish movement therefore began to raise issues of cultural autonomy for Flanders. In reaction to the gains of the Flemings, the small Walloon movement followed suit (Ruys, 1981, pp. 63-64). Works began appearing on Flanders and Wallonia, Fleming and Walloon congresses addressed issues of regional autonomy, and Flemish and Walloon activists began promoting the traditional symbols of nationalism: flags, songs, and slogans. Flanders and Wallonia were presented as natural territorial units, and the creation of regions as ideologies was underway.

Nevertheless, until World War I only the more radical fringes of the linguistic nationalist movements espoused and supported the idea of language regions as cultural and political units to which power should be devolved from the centralized Belgian state. During World War I, however, this view became significantly more widespread, particularly among Flemings. When the Germans occupied Belgium during the war, a small but significant number of the more radical Flemish nationalists agreed to cooperate with the Germans in exchange for vague promises of a degree of self-determination (Archives du Conseil de Flandre, 1929, p. 476). The Germans, seeking to capitalize on internal divisions within Belgium and to preserve the support of the Flemish activists, instituted an internal administrative division of the country in 1917 (Pirenne and Vauthier, 1926, p. 257). Laws were enacted creating separate administrative structures for Flanders and Wallonia and requiring the use of Dutch in Flanders in such fields as administration, justice, and education. Although these structures were immediately dismantled after the defeat of the Germans, their impact in fostering a regional view of Belgium based on linguistic criteria should not be underestimated. The formal creation of regions based on a particular criterion is one of the most effective ways for the criterion to take on lasting meaning as a basis for regional divisions.

Language conflicts within the Belgian army contributed to an increase in linguistically based regionalism as well. At the lower levels the soldiers were predominantly Flemish, while officers were overwhelmingly French-speaking (Clough, 1968, p. 212). Again, language was an element of differentiation. Officers gave orders in French and were not always understood by the soldiers. In a few highly publicized incidents the failure of the soldiers to understand commands had disastrous consequences. The significance of these circumstances for the development of regional consciousness is evident when one considers that the army brought together Flemings from all over the Dutch-speaking part of Belgium. While in the army this broad range of the Flemish population found that their language put them at a disadvantage. Soldiers from Limburg and their comrades from West Flanders came to realize that they had a common problem which united them. Out of this grew the Frontist movement, which also became a significant force in encouraging not only Flemish identity but the conception of Flanders as a regional entity (de Schaepdrijver and Charpentier, 1918).

Even though a period of suppression of the activist collaborators followed World War I, the view of Flanders as a region unto itself had firmly taken hold. The interwar years were characterized by a growing sentiment in northern Belgium of Flanders as a distinct cultural and territorial entity. This led to the enactment of the first truly territorial legislation in the 1930s. The new laws differed fundamentally from their predecessors by calling for unilingualism in Flanders in the fields of justice, administration, military service, and education (Maroy, 1966, pp. 460-473). This gave Dutch speakers a degree of control in their own region. In other words, these were laws based explicitly on a regional partitioning of Belgium along

linguistic lines which came about as a result of the
strategies of Flemish activists seeking, and I quote again
from Robert Sack's definition of territoriality, 'to affect,
influence, or control actions and interactions ... by
asserting and attempting to enforce control over a specific
geographical area' (Sack, 1981, p. 55).

The materials available from Walloon conferences
during this period indicate that in response to the legislative gains of the Flemings, certain Walloon leaders began
promoting the cause of Wallonia as a distinct cultural and
territorial unit (e.g. Congrès de Concentration Wallonne,
1930). This development was largely reactive in character,
however, and did not succeed in creating widespread regional
linguistic identification among the Walloon population. Yet
the very issues raised by the Walloon activists suggest the
strong territorial character which the Flemish movement had
assumed. These included the status of the Brussels region
as an increasingly Frenchified enclave within Flanders and
the implications of the new laws for Francophones living
either in communes along the linguistic boundary or in
Flanders itself (Schreurs, 1960, pp. 24-31).

Although the territorial legislation of the 1930s
promoted thinking about Belgium in terms of language
regions, the laws brought about only limited changes due to
widespread resistance by French speakers made possible in
part by the lack of enforcement mechanisms or sanctions.
Belgium on the eve of World War II was still strongly
unified from a functional standpoint even though many
Flemings and a few Walloons had come to attach considerable
importance to their language regions.

The occupation of Belgium by the Germans in World War
II did little to change the situation. There were some
collaborators on both sides and a post-war suppression of
the collaborators which was directed particularly against
the Flemings. During the 1950s, however, the economic
fortunes of the previously dominant industrial belt of
Wallonia declined rapidly. The downturn resulted from the
depletion of coal reserves and the decay of the industrial
infrastructure, as well as the growing attractiveness of the
North for new investment as a consequence of its locational
advantages and cheaper labour force (Quévit, 1978). At the
same time, the Flemish movement continued its pressure for
enforcement of the 1930s' language laws, a larger Flemish
role in state affairs, and increased unilingualism in
Flanders. The combination of these circumstances and
Flemish successes in a number of political and cultural
arenas created a growing recognition among Walloons of their
minority position within Belgium from a numerical, and
increasingly from an economic and political, standpoint.
Things came to a head with the general strikes of 1960-1961,
which were considerably more successful and widespread in

the South than in the North (Création d'un Mouvement Populaire Wallon, 1961). With territorial ideologies already reasonably well developed, and with the temporal and spatial association between economic change and Flemish ascendancy widely recognized, it is not surprising that for the Walloons the strikes took on a strongly territorial character. In a sense, the strikes functioned for the Walloons in a way somewhat analogous to the part played for the Flemings by military service in the Francophone-dominated Belgian army of World War I. People from many parts of the Walloon region were united by a shared economic grievance which had strong regional and linguistic overtones as a result of the successes of previous language related territorial initiatives. It is not surprising that this was the period during which Wallonia gained widespread acceptance as an important regional entity.

With territorial ideologies firmly entrenched among significant sectors of both the Flemish and Walloon populations, pressure mounted for major reforms of the Belgian state to give formal recognition to the language regions. The ambiguous status of Brussels in both the Flemish and Walloon conceptions of a revised state structure led in turn to a reactive movement among significant sectors of the Brussels population (Mols, 1961). This movement was, of necessity, strongly territorial in character. To make a long story short, the territorial emphases of the various so-called community movements led to the formal partitioning of Belgium along linguistic lines in the early 1960s. This was the beginning of a process of linguistic regionalization which culminated in two major revisions of the Belgian Constitution in 1970 and 1980. The new constitutional framework provides for a significant devolution of power to the institutions of the language regions and the division of a wide array of social, economic, and political matters along linguistic lines. Nevertheless, Belgium has stopped short of adopting a federalist structure (Stephenson, 1972), and regional institutions remain dependent on the central government for their budgets.

Although it is beyond the scope of the present study to analyze in detail the complexities of the new legislative and constitutional provisions, it is important to note that disputes of a territorial nature, which grew out of strongly felt regional ideologies, were substantially responsible for precipitating and furthering formal, linguistically based regionalization. These disputes included controversy over the territorial implications of a linguistic census, debate over the Frenchification of the area surrounding Brussels, and dissension over the presence of a French-speaking university (the Catholic University of Louvain-Leuven, Figure 9.1) in Flanders. Moreover, the effect of institutionalizing regional feelings through legal enactment has been to structure even more of the debate between the two language groups in territorial terms and to effectuate increased functional segregation of Flemings and Walloons. Since formal regional arrangements require fixed territorial delimitations, controversy has been focused on areas where there is a poor fit between language distribution and the legally demarcated regions. In addition, the splitting of institutions along linguistic/territorial lines has promoted debate on relative regional imbalances. Formal linguistic regionalism has thus served both to alter functional regions through the creation of discontinuities in patterns of spatial interaction and activity, and to change perceptions about place through the reification of regional ideologies.

Implications and Conclusions

While this brief overview of the Belgian case has required considerable generalization, the story of evolving linguistically based regionalism in Belgium provides insight into the importance for interethnic relations of the ways in which space is conceptualized and used. A few general points should be emphasized. First, the significance of language areas as regions has changed considerably during the course of Belgian history. To project present realities into the past by talking about Flanders and Wallonia as significant cultural and territorial units in the early years of Belgian independence makes little sense. Moreover, the growth, timing, and significance of ethnoregional consciousness cannot be explained solely by reference to structural factors such as large-scale cultural or economic inequalities. Differentiation along these lines did not entirely correspond to language patterns, and did not always translate into regional consciousness. Rather, these objective or structural factors must be seen as elements of the context in which ethnoregional consciousness arises (Murphy, 1985), with an explanation of actual developments grounded in historical circumstances and structured behavioural choice (Agnew, 1981). In this regard, we must be careful not to allow the pervasiveness of present regional conceptions to lead us to assume that these conceptions are useful in reconstructing past inequalities, for the choice of region can obscure subregional or intraregional differences which, at the time of inquiry, could have been as important as the inequalities between the regions which eventually emerged.

In a different vein, the Belgian experience with regionalism demonstrates the importance of territoriality as a force in shaping the nature and extent of intergroup tensions. Territorial divisions are not simply the by-product of linguistic differences which can be treated as passive elements in our inquiries into the social and political transformations which lead to the development of substate nationalism. Rather, territories, as human constructs, can serve to change social relations and to redefine the nature and scope of social problems. For Belgium, once the territorial issue became an important priority of the Flemish movement, the focus of debate shifted. For example, the preservation of the unilingual character of the area surrounding Brussels became as important an issue as guaranteeing the right of Dutch speakers to use their language in court. Even more strikingly, the types of issues currently being debated are direct outgrowths of the institutionalization of territorial strategies through the legislative creation of formal linguistic regions. These include the existence of a Francophone minority in Flanders, the inclusion of the Fourons area in the Flemish region, and the containment of the spread of French in the area surrounding Brussels (McRae, 1986, pp. 275-322). The nature and extent of each of these problems is fundamentally tied to the conceptual and formal regionalization of Belgium.

Finally, examination of the ways in which space is conceptualized and used is an important analytical tool in understanding the Belgian case. It draws attention to the development of regional consciousness within ethnolinguistic groups, the significance of territorial strategies for intragroup and intergroup relations, the causes and consequences of overlapping senses of territory among groups, and the relationships between formal, institutional regionalization on the one hand, and human spatial organization and perception on the other. In the September, 1985 issue of Progress in Human Geography Colin Williams points to the place in geography for the study of territorial issues of this sort (Williams, 1985, p. 340). He notes:

> [W]e know very little in detail of the manner in which nationalist elites have succeeded in cementing the relationships between the formation of a nation's identity and a specific locale. Much work remains to be done in tracing the historical manipulation of claims and counterclaims to specific homelands and on the redefinition of group rights to territory. ... Despite earlier work on culture areas, national frontiers and homeland development, few geographers have addressed themselves to an explicit analysis of the way in which political leaders have identified and channeled group expectations to territorial control, a fascinating and worthy project to study.

While we need not limit ourselves in our research to the role of political leaders, the case for a focus on territorial perceptions and strategies is well stated by Williams. The Belgian situation provides an excellent example of the importance of this approach.

Acknowledgement

This research was supported in part by a Fulbright-Hays grant to Belgium in 1985-6.

References

Agnew, John A. (1981) 'Structural and Dialectical Theories of Political Regionalism.' in Political Studies from Spatial Perspectives. Edited by Alan D. Burnett and Peter J. Taylor. New York: John Wiley & Sons, pp. 275-289

Alverson, Hoyt S. (1979) 'The Roots of Time: A Comment on Utilitarian and Primordial Sentiments in Ethnic Identification.' In Ethnic Autonomy: Comparative Dynamics, the Americas, Europe, and the Developing World. Edited by Raymond L. Hall. New York: Pergamon Press, pp. 13-17

Annuaire Statistique de la Belgique (1870) Brussels: Ministere de l'Intérieur

Archives du Conseil de Flandre. (1929) Brussels: Dewarichet

Clough, Shepard B. (1968) A History of the Flemish Movement in Belgium - A Study in Nationalism. New York: Octagon Books

Congrès de Concentration Wallonne - Liège 27-28 Septembre 1930 - Compte-Rendu Officiel.(1930) Liège: n.p.

'Creation d'un Movement Populaire Wallon' (1961) Courrier Hebdomadaire du CRISP, 101, March 24

De Schryver, Reginald (1981) 'The Belgian Revolution and the Emergence of Belgium's Biculturalism' In Conflict and Coexistence in Belgium - The Dynamics of a Culturally Divided Society. Edited by Arend Lijphart. Berkeley: Institute of International Studies, pp. 13-33

Dickson, Tim (1986) 'High Cost of Cultural Divide.' Financial Times, June 13

Dominion, Leon. (1917) The Frontiers of Language and Nationality in Europe. New York: American Geographical Society

Fredericq, Paul (1906) Geschiedenis der Vlaamsche Beweging. Ghent: Vuylsteke

Ginsburg, Norton S. and Chester F. Roberts, Jr. (1958) Malaya. Seattle: University of Washington Press

Gysseling, Maurits. (1975) 'Vlaanderen (Etymologie en Betekenisevolutie).' In Encyclopedie van de Vlaamse Beweging. Tielt: Lannoo, pp. 1906-1912

Hechter, Michael. (1975) Internal Colonialism: The Celtic Fringe in British National Development. Berkeley: University of California Press

Henry, Albert. (1974) Esquisse d'une Histoire des Mots 'Wallon' et 'Wallonie'. Brussels: La Renaissance du Livre

Knight, David B. (1982) 'Identity and Territory: Geographical Perspectives on Nationalism and Regionalism.' Annals of the Association of American Geographers, 72, pp. 514-531

Luykx, Theo. (1959) Atlas Historique et Culturel de la Belgique. Brussels: Elsevier

Maroy, Pierre (1966) 'L'Evolution de la Législation Linguistique Belge' Revue du Droit et de la Science Politique, 82, pp. 449-501

McRae, Kenneth D. (1983) Conflict and Compromise in Multilingual Societies, Volume 1, Switzerland. Waterloo, Ontario: Wilfried Laurier Press

_____. (1986): Conflict and Compromise in Multilingual Societies, Volume 2, Belgium. Waterloo, Ontario: Wilfrid Laurier Press

Mikesell, Marvin W. (1983): 'The Myth of the Nation State.' Journal of Geography, November-December, pp. 257-260

Mols, Roger. (1961): Bruxelles et les Bruxellois. Louvain: Société d'Etudes Morales, Sociales et Juridiques

Murphy, Alexander B. (1985): 'Partitioning as a Response to Cultural Conflict.' Geographical Perspectives, 55, Spring, pp. 53-59

Nairn, Tom. (1977): The Break-up of Britain. London: New Left Books

Orridge, Andrew W. and Colin H. Williams. (1982): 'Autonomist Nationalism: A Theoretical Framework for Spatial Variation its Genesis and Development.' Political Geography Quarterly, 1, pp. 19-39

Pirenne, Jacques and Marcel Vauthier. (1926): La Législation et l'Administration Allemande en Belgique. New Haven: Yale University Press

Polasky, Janet. (1981): 'Liberalism and Biculturalism.' In Conflict and Coexistence in Belgium - The Dynamics of a Culturally Divided Society. Edited by Arend Lijphart. Berkeley: Institute of International Studies, pp. 34-45

Quévit, Michel. (1978) Les Causes de Déclin Wallon. Brussels: Vie Ouvrière.

Ruys, Manu. (1981) The Flemings - A People on the Move, a Nation in Being. Tielt: Lannoo

Sack, Robert. (1981): 'Territorial Bases of Power.' In Political Studies from Spatial Perspectives. Edited by

Alan D. Burnett and Peter J. Taylor. New York: John Wiley & Sons, pp. 53-71

_____. (1983): 'Human Territoriality: A Theory.' <u>Annals of the Association of American Geographers</u>, 73, pp. 55-74

de Schaepdrijver, Karel and Julius Charpentier. (1918): <u>Ontwikkelingsgang der Vlaamsche Frontbeweging.</u> Brussels: De Raymaeker en Peremans

Schreurs, Fernand. (1960): <u>Les Congrès du Rassemblement Wallon de 1830 à 1959.</u> Charleroi: Institut Jules Destree

Senelle, Robert. (1978) <u>The Reform of the Belgian State.</u> Memo from Belgium No. 179. Brussels: Ministry of Foreign Affairs, External Trade and Cooperation in Development

Smith, Anthony D. (1982): 'Nationalism, Ethnic Separatism and the Intelligentsia.' In <u>National Separatism.</u> Edited by Colin H. Williams. Vancouver: University of British Columbia Press, pp. 17-41

Stengers, Jean. (1981): 'Belgian National Sentiments,' In <u>Conflict and Coexistence in Belgium - The Dynamics of a Culturally Divided Society.</u> Edited by Arend Lijphart. Berkeley: Institute of International Studies, pp. 46-60

Stephenson, Glenn V. (1972): 'Cultural Regionalism and the Unitary State Idea in Belgium.' <u>The Geographical Review</u>, 62, 4 October, pp. 501-523

Verdoodt, Albert. (1977): <u>Les Problèmes des Groupes Linguistiques en Belgique.</u> Louvain: Institut de Linguistique

_____. (1983): <u>Bibliographie sur le Problème Linguistique Belge.</u> Quebec: Centre International de Recherche sur le Bilinguisme

Williams, Colin H. (1985): 'Conceived in Bondage - Called unto Liberty: Reflections on Nationalism.' <u>Progress in Human Geography,</u> 9, pp. 331-355.

Zolberg, Aristide R. (1974): 'The Making of Flemings and Walloons: Belgium: 1830-1914.' <u>Journal of Interdisciplinary History</u>, 5, pp. 179-235

Chapter 10

NATIONALISM, SOCIAL THEORY AND THE ISRAELI/PALESTINIAN CASE

Juval Portugali

This study results from an attempt to build a theoretical
framework for a study of Israeli-Palestinian relationships.
In this attempt we have first surveyed the scientific
literature concerning these relationships (Newman and
Portugali, 1987) and found that the few studies which employ
and discuss theoretical models are derived almost exclusively from recent social theory. In contradiction to
emphases in the 1950s and 1960s on notions of modernisation
and nation-building, recent social theory tends to interpret
social conflicts in terms of ethno-economic antagonism and
to conceive of capitalism as the driving force of social
relations.

Thus, the 'queuing competition model', Marx's (1970)
notion of the Industrial Reserve Army, Bonacich's (1972)
'split labour market theory', the 'core-periphery model',
Horowitz's (1982) dual-authority-policy model, Sandler and
Frisch's (1984) multi-centre periphery model, Whyte's (1978)
'external versus internal conflict approaches', Smooha's
(1978) conflictual-neo-pluralistic approach, ... , all have
been applied to the Israeli Palestinian case (for a detailed
bibliography see Newman and Portugali, 1987).

Indeed, much insight has already been gained by
applying the various facets of social theory to analysis of
the evolving relations between Israelis and Palestinians.
Yet, the attempt to explain these relations in terms of
ethno-economic conflict in a plural society, or in terms of
capitalist class relations, has been only partial and
limited in its success. At a more subtle level and in a
wider historical perspective these interpretations fail to
take into account some of the most elementary facts. Thus,
for example, an interpretation of the relations in terms of
ethnic conflict in a plural society obscures the fact that
one is observing a nationalistic struggle for national
self-determination. Similarly, interpretations in terms of
class conflict in a capitalist society overlook the fact
that war and state action were not the consequence of class

relations, but rather the starting point and framework for their development (Portugali, 1986, p. 368). Also, they overlook the historical scenario in which the activities of nationalist movements have provided the foundation for the emergence of class relations between Israeli Jews and Palestinian Arabs.

The crux of the matter is the persistence of social theories to preserve an anachronistic conceptual separation of state, civil society and economy (Frankel, 1982); its causal-mechanistic separation between the material and spiritual domains; and its disregard of socio-spatial relations and consequently its misconception, or under-estimation, of the role of nationalism in the existing social order.

In the following I shall elaborate on the above points and suggest a theoretical framework for the interpretation of nationalism and its role in the Israeli-Palestinian case.

Two Types of Social Order

In a previous paper (Portugali, 1985) I have suggested interpreting social theory as emerging and developing out of the dialectical tension created by inconsistencies between the mechanistic-cartesian order, which dominates science, and the apparent property of society as an undivided whole. In contrast to positivist social theory which conceives the above inconsistencies as temporary, social theorists such as Marx, Durkheim, and recently, Levi-Strauss, Althusser, Giddens and the constructivist school, suggest taking them seriously. However, recent social theory, as derived from the above theories, is in several aspects still mechanistic, or mechanistic in disguise: firstly, since it assumes the human domain as essentially independent of the natural domain; secondly, and specifically relevant for the present discussions, for assuming the essential separation, and consequently casual relations, between the material and spiritual domains of human reality.

This is particularly prominent, for example, with Marx's base and superstructure and the postulation that the individual's spiritual domain is determined by material reality. (In this connection see Althusser's (1969) conception of 'economic determination in the last analy-sis'.) This is also found in Giddens (1981) who, with his notion of 'structuration', has gone far beyond crude 'structural determinism'. However, while in Giddens' 'structuration' theory, human agency and society enfold each other, he still insists on separating the material and spiritual. With respect to the present discussion he insists on separating nationalism from nation-state: the first being a psychological-spiritual-sentimental phenomenon and the second, material-institutional (Giddens, 1981, pp. 190-196).

Such a mechanistic separation between the material and spiritual not only contradicts the reality of the Israeli-Palestinian situation with respect to nationalism, but is also conceptually erroneous and unnecessary (Portugali, 1985). As an alternative I present below the notions of 'implicate social order' and 'generative social order'.

The implicate social order

The notion of implicate social order, proposed here for the analysis of nationalism in the Israeli-Palestinian confrontation, was inspired mostly and directly from Bohm's (1980) philosophy regarding the theory of implicate order, as well as from Prigogine's (1981) notion of self-organisation and Haken's (1984, 1985) 'synergetics'. The connection between these theories and social theory has been described elsewhere (Portugali, 1985). Here I shall list several properties of the implicate social order which will facilitate the discussion of nationalism that follows.

First is the notion that society (indeed reality) must be conceived of as an undivided whole in an ever-evolving movement. From this standpoint, social entities such as nation, class, and, at a deeper level, the human being itself are not 'building blocks' independent of each other, but 'social events' - inseparable from each other and from their wider spatio-historical context. They are inseparable in the sense that each social entity <u>enfolds</u> all other entities and society as a whole. This property I have termed <u>implicate social relation</u> and have applied it to Israeli and Palestinian societies in a study of Arab labour in Tel-Aviv (Portugali, 1986). The working hypotheses, which emerge out of this study, indicate that Israeli and Palestinian entities enfold each other to the extent that neither is definable today independent of the other. From the turn of the century this was a characteristic specifically of the ideological-superstructural level. However, since 1967 this process of enfoldment entered a new phase by including the infrastructural domain of everyday spatial and socio-economic interaction between Israelis and Palestinians. This view of individual-society relations is close to Marx's 'relational method as interpreted by Harvey (1982), and also to Giddens' (1981) notion of structuration.

Second, the notion of implicate relations applies also to the distinction between the material (base) and spiritual (superstructure) domains. They are both conceived of as 'social events', which with relative autonomy in time and space enfold each other as abstractions from social reality. Thus both nationalism as a psychological, superstructural

phenomenon and the nation-state as a material, territo-
rial-institutional entity, are seen as abstractions from a
deeper implicate social domain in which they merge as
inseparable.

Third, the notion of implicate relations refers also
to the distinction between 'past' and 'present'. Thus
present social entities such as nationalism or nation-state
(or class and ethnic groups ...) not only enfold each other
but also their past. They are explications or abstractions
from past realities (This conception is close to Giddens'
notion of 'world time').

In order to grasp better the above properties I shall
introduce Bohm's notion of 'generative order' which is close
to Haken's concept of 'order parameter'.

The generative social order

It may be convenient to begin by considering mechanistic and
holistic conceptions of change. In the first, equilibrium
is seen as the natural state of reality and change as a
causal disturbance in an otherwise stable society. One is
here dealing with causal relations and change. Holistic
conceptions on the other hand start from the postulation
that there is movement and disequilibrium and consequently
what must be explained is how stability is created and
terminated. Here entities are seen as spatially and
temporally confined formal appearances, or events.

Stability is achieved when a given configuration of
events predominate. The amplitude of this configuration is
termed 'generative order' by Bohm (personal communication)
and 'order parameter' by Haken (1985, p. 207). Once a given
order parameter is established, other events or configurations 'are subjected to the newly established order state
or, they are "slaved) by the order parameter' (Haken, 1985,
p. 207). As interpreted below, nationalism became the
generative order in modern society, first, in the sense that
world society was, and still is, 'forced' to organise
according to its order principles and, second, in the sense
that other major social orders, such as capitalism and
communism, accept (i.e. were enslaved by) its principles.

This process of 'enslavement' can be described in
terms of innovation diffusion in space-time, i.e. the
diffusion of a certain order parameter (innovation) from its
point of origin. This is exemplified below in the discussion of the space-time diffusion of nationalism from its
European core to Zionism and later to Palestinianism.

A given social configuration (i.e. social order) has a
material content and an informational content. Thus
nationalism can be seen as a generative social order, with
the nation-state as its material content (territorial,

institutional ...) and the ideology of nationalism as its informational content.

The notion of an informational content of a generative social order (whose origin is in a system way of thinking) is associated with the notion of 'ideology' and in the double nature of the term: ideology as a 'plan' of what should be, and ideology as 'false consciousness' which obscures human vision from the real conditions of existence (Larrain, 1982). Thus, nationalism is simultaneously a 'plan' for society's geopolitical organisation and a false consciousness as it presents social constructs such as territorial homeland, nation and nation-state as 'natural' and 'eternal'. Both properties of the ideology of nationalism are best illustrated by its 'core doctrine'. Briefly the doctrine holds that:

1.   Humanity is naturally divided into nations.
2.   Nations are known by certain characteristics which can be ascertained.
3.   The only legitimate type of government is a national self-government.
4.   The primary condition of global freedom and harmony is the strengthening of the nation-state.
5.   For freedom and self-realisation, people must identify with a nation.
6.   Loyalty to the nation-state overrides other loyalties.
7.   The nation-state's supreme and sole obligation is towards its co-nationals.
8.   Nations can only be fulfilled in their own territory, with their own state and government.
9.   The nation-state - the unity of people, territory and government - is the genuine unit within and through which people conduct their social, economic and cultural affairs.
[Properties 1-3 are from Kedourie (1960, p. 1); properties 4-16 from Smith (1971, p. 1) and properties 7-9 from Portugali (1976).]

This list of propositions, referred to above as 'the core doctrine' of nationalism, is also a general description of present world society - a world divided into nation-states each striving to protect its co-nationals politically, culturally and economically: 'But what now seems natural was once unfamiliar, needing argument, persuasion, evidence of many kinds' (Kedourie, 1960, p. 1).

Using the above conceptual framework we can now turn to an interpretation of nationalism in general and its connection to the Israeli-Palestinian case in particular.

The Space-Time Diffusion of the Nationalist Generative Order

In his 'dual legitimacy' theory of nationalism Smith (1971, pp. 231-6) suggests the following historical sequence for the emergence of nationalism and nation-state:

Empire/Possessive state ⟶ Scientific state ⟶ Nation state

The notion of a scientific state is connected with the scientific world view, with whose rise the question of the source of cosmic and social order became problematic. At the beginning, with the rise of the Cartesian-Newtonian mechanistic world view, God was still retained as the source of cosmic and social order. However, the consequent evolutionary paradigm implied that reference could no longer be made to God as the integrative authority of social order, but to 'natural', 'scientific', 'objective', 'observable' socio-cultural traits, such as language, ethnic origin, common religion or socio-economic classes. The social elites, the rules of the 'scientific state' and social theory, have referred to these as legitimate, integrative criteria, and to the evolutionary struggle for survival as the motor of human history: for liberals, a struggle between selfish individuals in a perfectly competitive market; for marxists, a class struggle; and for Spencerians, a struggle between nations.

The result was to delegitimise the social groups who did not belong to the dominant political-cultural entity (i.e. groups who could not fit or conform with the newly emerging generative social order) and make them conscious of their ethno-cultural uniqueness (Smith, 1971). Such minority groups were confronted with a crisis: either to assimilate in the new social entity by giving up their cultural heritage (and they were not always welcomed by the recipient group), or to preserve their cultural tradition by conforming with the new generative order, that is, by claiming and struggling for their own national self-determination, with their own territory, government and state. In Haken's terminology, they had no choice but to be enslaved by the newly emerging order parameter.

The space-time diffusion of nationalism and the nation-state from their European origin is thus a process of enslavement, as more and more socio-cultural groups are being exposed to nationalism as a generative order, first in Europe and later in its colonies. They were 'exposed', firstly, to the informational content of nationalism, and secondly to the dialectics of space. Several societies, such as the Germans, the Palestinians and post-colonial nation states, became conscious of their national identity after an external nation-state had defined them as a spatio-political entity. In this process the various socio-cultural groups responded by bringing to the fore socio-cultural traits (such as common language, territory, religion; common enemy, exploiter, ruler etc.) which 'scientifically' prove the existence of their nation as a natural-legitimate entity. Thus the variety of types of nationalisms, identified and typologised, for example, by Smith (1971) and Portugali (1976), reflect different groups' pasts or their responses to the advancing nationalist generative order. The rise of Zionism and its connection with the emergence of Palestinianism is one of the finest examples of this process of diffusion. ˉ

Zionism and Palestinianism

Within the nineteenth century European reality of emerging nation-states, the Jews (especially their intelligentsia) very quickly found that the strategy of assimilation was not an easy task. Jews were rejected by all sections of society, be they middle-class intelligentsia or social-ist-revolutionary movements. The personal history of many early Zionist leaders is probably the best illustration here. People like Herzl, Borochov and others emerged with Jewish nationalism (i.e. Zionism) as a consequence of disillusionment from the strategy of assimilation. Jews were reminded, time and again, that seemingly 'modern', 'rational', 'scientific' societies, be they liberal, socialist or Marxist-internationalist, all enfolded their past conception of Jews; that the enlightened, modern, rational, social principles are but a thin and fragile surface covering an ocean of prejudice and hatred. The Dreyfus affair, the Pogroms that followed the failure of the 1905 revolution, the Nazis ..., are but few examples from a long list.

Thus, despite their global spatial dispersion and their deep involvement in their countries of residence, the Jews were extracted from their native countries in order to be enslaved by nationalism - the emerging generative social order - that is, to be conscious of, and develop, their own 'natural' ethno-national distinctiveness, their own 'nat-ural' territory and their own 'scientific' self-government.

Jewish nationalism exposes and emphasises one of the major components of nationalism which is often overlooked in scientific discussions, that is, the territorial component. Within the European context the Jews had all of the neces-sary criteria to self-determine themselves as a nation (language, religion, ethnic origin, history, culture ...), but one, territory. Thus emerged the 'Jewish Problem' of a nation without territory, and thus the name Zionism - the name of the ancient homeland - and so Smith (1971) typolog-ises Zionism as a 'diaspora movement'. A national territory became the central goal of Zionism, even for socialists and Marxists like Borochov, who considered a Jewish national homeland as the precondition to revolutionise the Jewish people socio-economically.

Territory and self government, were not to be found in
the Jewish European 'present', and the Jews had to turn to
their historical past. In this respect the Zionist movement
simultaneously enfolded its 'present' European social
context and its 'past' - an explication of (or abstraction
from) European social reality and Jewish past history.
The Zionists who migrated to Palestine were probably
the main agents to transmit or diffuse spatially the nationalist generative order from its European origin to Palestine
(the other route was through the Young Turks revolution).
Their successful activities in establishing a new 'healthy',
'modern', 'scientific', Jewish society, with its own
language, own settlement, production, health, security and
judicial systems in their ancient homeland, have acted on
the local Arab population in a way similar to the impact of
European nationalism on the Jews: it spatially disconnected
the local Arab population, who lived in the territory
claimed by Zionists, from other Arab populations and thus
made them conscious of their own cultural uniqueness, their
own common fate, their own tradition. Just as the Zionist
hold in Israel became firmer and more elaborate, so the
Arabs in the same territory became a nation which, like the
nineteenth century Jews, had no choice but to be enslaved by
the 'only' legitimate medium to claim their rights: by
nationalism as a generative order.
Palestinian national identity, in this respect, is to
a large extent a Zionist creation. Thus, from its very
origin Palestinian identity enfolded Zionism. Yet, simultaneously, this emerging Palestinian entity has transformed
and shaped the evolving Israeli society. As a 'diaspora-
ethnic' nationalist movement (Smith, 1971; Portugali, 1976),
Zionism has striven not only to establish a Jewish nation-state in Palestine but also to 'proletarianise',
'ruralise' and thus 'normalise' Jewish society.   These
national goals, originating as they did in the Judaeo-
European social reality, became conditional for their
implementation upon, and consequently redefined to take
account of, the Palestinian Arabs. 'Proletarianisation' was
transformed into 'the conquest of labour' (from Arab
labourers), 'ruralisation' into the 'conquest of land' (from
the Arabs), 'Hashomer' - a Jewish guard organization - in
place of, and against, the Arabs. Zionist and Palestinian
societies thus 'enfolded' and 'contained' each other to the
extent that neither could be defined exclusively independent
of the other.
This applies also to the recent wave of Palestinian
nationalism under the leadership of the Palestinian Liberation Organisation. The latter can be regarded as if not a
Zionist at least an Israeli creation.   Here also the
dialectics of space were crucial; particularly with respect
to the Israeli conquest of the West Bank and Gaza Strip in

1967 and their spatio-political disconnection from Jordan and Egypt.

Up to 1967, especially between 1948 and 1967, one could observe a process of externalisation of the Israeli-Palestinian conflict, in the sense that the majority of Palestinians lived outside Israel and under Arabic rule (mainly Egyptian and Jordanian). While under Egyptian rule the Palestinians were kept concentrated and isolated in their refugee camps: in Jordan they underwent a process of integration, forming a national-ethnic majority ruled by a minority of Bedouins.

The Israeli conquest in 1967 spatially disconnected the West Bank Palestinians from their Hashemite rulers and the Gaza Strip Palestinians from their Egyptian rulers. This has spatially united the West Bank, Gaza Strip and Israeli Palestinians and, finally, achieved proletarianisation of the Palestinian society by the spatial and economic integration of Israel and the occupied territories (Portugali, 1986). This proletarianisation process has transformed the Israeli-Palestinian conflict from an ethno-nationalist into an ethno-nationalist-class conflict. Israel became the direct ruler of the Palestinians in the political-military sense, as well as in the socio-economic sense. This new reality provided the foundations for the rebirth of the Palestinian national movement under the leadership of the PLO, as well as its new leftist orientation.

Nationalism and Social Theory

As interpreted above, nationalism as a generative social order arose (with modern, industrial-scientific society) out of the dialectical tension between the emerging mechanistic-scientific world view, on the one hand, and the holistic social reality and world view which predominated in pre-industrial European society, on the other hand. The same applies to social theory as a whole and, in fact, to other major political ideologies which arose parallel to the rise of nationalism, that is, to Marxism-socialism-communism and to liberalism-capitalism. These too can be regarded as different configurations of social events, or as 'competing' generative social orders. However, unlike these competing generative orders which subscribe to either atomism-individualism or holism-internationalism, nationalism seems to synthesise the two elements. In contradiction to the liberal vision of society of individuals operating under the guidance of the invisible hand of a perfectly competitive market, and in contradiction to Marxists' vision of an international classless society, nationalists' vision seems to satisfy simultaneously the mechanistic-scientific imperative for division and the holistic conception of social unity. Society's natural building blocks are nations; each with its own territory, upon which, in order to fulfil and reproduce itself, the nation must establish its own nation-state. Each nation is thus a unity of people, territory and state.

Nationalism is the only generative order without an (or with the least dominant) associated body of scientific theory. Liberalism-capitalism has, for example, its classical political economy and neo-classical economics, while socialism-communism has its Marxist theory. Yet, as a generative order, nationalism dominates or enslaves these other social configurations in the sense that it is the only social order parameter which is commonly accepted by communists and capitalists alike. This property is also enfolded in modern social theory. The latter is composed mainly of two rival standpoints (Marxism vs Liberalism) which despite essential differences also share several characteristics. First, as noted, both accept the core doctrine of nationalism, though for different reasons and argumentations: Marxist-Leninists, as a temporary necessary stage until the nation-state will wither away by itself; Liberals as an organizational framework to supply the socially needed public goods, in face of a 'market failure' (Bator, 1958).

Second, both theories (and thus social theory as a whole) are essentially aspatial or ageographical. Indeed, the recognition of this has provided the foundation and rationale for modern human geography. The latter commenced with the criticism that economic and sociological theories were developed as if social and economic processes took place 'on a head of a pin'. This also provides the ground and rationale for recent Marxist geography in response to the fact that 'Marx generally theorizes about Capitalism as a "closed economic system" ... (Harvey, 1982, p. 413) ... of preexisting ... nation states rather than of the processes that give rise to spatial configurations in the first place' (Harvey, 1985, p. 45). Recently, Giddens (1981) has also become aware of this fact, and his social theory was indeed inspired by human geography. However, the consequence of this fact was that the question of what is, or what should be, the proper geopolitical unit to enjoy self-government was either left out of the discussion or overlooked.

Third, both are essentially ahistorico-traditional, in the sense that liberalism and classical economic theory proceed from the assumption of rational economic individuals and Marxism from the notion of 'historical stages' and social classes: Marx and Engels adopted Morgan's social evolutionary scheme in which society is seen as developing in stages, where each sequential stage destroyed its previous. The consequence was that Marx and Engels did not fully appreciate the role of traditional-cultural traits in shaping future society. This is a point on which social-ists, such as Hess, Kropotkin, and recently Buber, have departed from, and criticised, Marx (see Buber, 1983).

The combined consequence of the second and third properties was that the question of what should be the proper socio-cultural and political-spatial unit to enjoy self-government could never be posed by social theory. Furthermore, because of their ahistorical/cultural property, both liberalism and Marxism (and consequently social theory) ignore the way past social entities enter into present reality. Again, Giddens and his notion of 'world time' seems to be a start in the right direction.

Nationalism as a generative order and political ideology seems to provide a clear-cut answer to the above questions on which other political ideologies remain silent. More than other potential generative orders, nationalism is an explication of past history. It takes already existing socio-cultural entities (ethnic/national group, state, science) and reorganises their interrelationships as the foundation for social order - unlike Marxism/liberalism which almost ignore their existence and role in creating future society. Hence, while Marxism/liberalism might rightly claim to be more 'rational' or 'scientific' in their conception of society, nationalism is more 'realistic'.

Yet, nationalism does not furnish a complete theory of political action and says little about the internal struc-ture of society. In this respect it has complementary relations with other political ideologies which, from the point of view of nationalism, can be defined as 'supporting ideologies' (Portugali, 1976). They are supportive in two respects. Firstly, the nationalist principle of self-deter-mination also implies that nations should self-determine their internal socio-economic-political order, but not what it should be. Liberalism, capitalism, socialism and Marxism support nationalism by suggesting alternative schemes regarding the internal structure of a national society. As noted, Marx developed his theory by treating human society as a closed system nation-state.

Secondly, the supporting ideologies are supportive also with respect to the 'false consciousness' property of an ideology. Thus, in Marxist theory the nationalist vision of society, vertically divided into nations, is rightly defined as ideological since it obscures the 'scientific', horizontal, division of society into social classes. Yet, the above discussion implies that this also goes the other way round. The fact that social theory is derived mostly from liberalism/Marxism implies that social reality is not interpreted in space/time terms, thus obscuring the exis-tence, role and importance of nationalism as a generative order. In this respect, nationalism as a generative order enslaves liberalism/Marxism and at the same time hides and flourishes behind their propaganda. Thus outright nationalist struggles for national-territorial self determination are interpreted by both liberals and Marxists in terms of liberation or class struggle and thus justify the greatest crimes against humanity.

Such complex relations between nationalism and its supporting ideologies also typify Palestinianism and Zionism. As for the latter, from its very origin all ideological streams were present in the Zionist movement, ranging from a minority of right-wing revisionists, to liberals, socialists and Marxists, who dominated the movement. For liberals like Herzl as well as for Marxists like Borochov, a Jewish state was not an end in itself, but the means to solve the 'Jewish problem'. Thus, in Herzl's Altneuland the Jewish problem must be solved by establishing a liberal-democratic capitalist state, very much in line with the West European model, while Borochov (1955), in his Our Platform, applies the Marxist principles of dialectical materialism to define the 'Jewish problem' as that of people without territory, peasantry, working class ..., and consequently to define Zionism as the means to revolutionise the Jewish people. Zionism was thus advancing behind and shadowed by its supporting ideologies as well as its liberal and, especially, socialist achievements. The Kibbutz, the Moshave, and the Histadruth are among its better known socialist innovations. Yet, the consensual foundation for the coexistence of liberalism, revisionism, socialism and Marxism was the Zionist generative order. And this nationalist generative order - this precondition to revolutionise socially the Jewish people - was also the ideological justification for the 'unavoidable, necessary evil' of the expulsion of Arab peasants from their land, for excluding them from the social achievements of Zionism, for making them refugees, for settlement activities and military control in occupied territories.

The same with Palestinianism: in the name of social justice, socialism, liberation and equality, generations of Palestinians are kept in refugee camps, not by Israelis or other 'enemies of the nation', but rather by their very leaders and co-nationals. Thus, up to 1967, the Egyptian authorities in the Gaza strip and the Jordanian authorities in the West Bank made sure that Palestinian refugees would not be able to improve their own personal welfare, or to leave the camps, or find a job. This was with the aid of the United Nations and various human rights organizations, which through UNRA have supplied the camps with a permanent flow of free food and other welfare facilities, everything but employment or productive activities. Every effort was made to ensure that the Palestinian person will not stand on his/her own feet, every effort to keep the Palestinians in the camps in a state of personal dependency and human degeneration. Since 1967 the paradox is even greater, for the Israeli authorities are striving to improve the situation in the camps by supplying jobs, better welfare and housing (not from altruism, of course, but as part of an attempt to dissolve the conditions for terrorism). Yet the major opposition to improve the conditions in the camps comes from the PLO who plainly threaten the life of 'collaborators' and from the Palestinian scholars like Abu-Ayyash (1976) who, within the frame of his 'core-periphery' interpretation of the Israeli planning activities in the occupied territories, condemns the Israeli authorities for their attempt at 'dismantlement of the Palestinian refugee camps and the development of an efficient form of transportation.' Here also the nationalist order parameter is advancing behind the false consciousness created by the enslaved ideologies of Marxism and liberalism and by social theory itself.

The Moloch

Nationalism as a generative social order is the Moloch to which both Zionists and Palestinians worship and by which both people are united. It is the nationalist generative order which has created the modern form of both nations, and their activities in war, terror, expropriation, are derived from this very source. It is for the idol of nationalism that both nations sacrifice their best children; it is for this Moloch that Palestinian 'fighters' have massacred Israeli schoolboys and girls in Ma'alot, and the Israeli army has bombarded the civilian quarters of West Beirut. It is the Moloch of nationalism as a generative order which is advancing behind the false consciousness created by liberal, socialist or Marxist ideologies, and by social theory itself. And this is particularly so when social theory is employed in scientific discussions of the Israeli-Palestinian conflict.

Scientists studying Israeli- Palestinian relationships must be aware that, whether they like it or not, they are participants and not observers. And if there is a scientific tool which might be constructive here, it is the property of criticism and self-criticism. With this property science can expose the fact that the Moloch of nationalism is not God's creation nor Nature's creation, but a human creation; that humanity is not naturally or divinely divided into nations, that nations do not have natural territories and that the state is not the sole precondition for human freedom.

References

Abu-Ayyash, A. (1976) 'Israeli Regional Planning Policy in the Arab Occupied Territories', Journal of Palestinian Studies, 5, no 4/5, 83-108.

Althusser, L. (1969) For Marx, Harmondsworth; Penguin.

Bator, F.M. (1958) 'The anatomy of market failure', Quarterly Journal of Economics.

Bohm, D. (1980) Wholeness and the Implicate Order, Routledge and Kegan Paul, London.

Bonacich, R. (1972) 'A Theory of Ethnic Antagonism': the split labour market', American Sociological Review, 37, 547-559.

Borochov, B. (1955) Ctavim vol. 1, Tel Aviv, 1955, Hebrew.

Buber, M. (1983) Pathes in Utopia, Am Oved, Tel Aviv, in Hebrew.

Frankel, B. (1982) Beyond the State?, Macmillan, London.

Giddens, A. (1981) A Contemporary Critique of Historical Materialism, Macmillan, London.

Haken, H. (1984) The Science of Structure: Synergetics. Van Nostrand Reinhold, New York.

Haken, H. (1985) 'Synergetics - An Interdisciplinary Approach to Phenomena of Self-Organisation', in Portugali J. (ed.) Links Between the Natural and Social Sciences, Geoforum, special theme issue, 16, 205-212.

Harvey, D. (1982) The Limits of Capital, Basil Blackwell, Oxford.

Harvey, D. (1985) The Urbanization of Capital, Basil Blackwell, Oxford.

Horowitz, D. (1982) 'Dual Authority Politics', Comparative Politics, 14,

Kedourie, E. (1960) Nationalism, London.

Larrain, J. (1982) The Concept of Ideology, Hutchinson, London.

Marx, K. (1970) Capital Vol. 1, Lawrence and Wishart, London.

Newman, D. and Portugali, J. (1987) 'Israeli-Palestinian Relations as Reflected in the Scientific Literature', Progress in Human Geography, 11.

Portugali, J. (1976) 'The Effect of Nationalism on the Settlement Pattern of Israel', unpublished PhD thesis, London School of Economics and Political Science.

Portugali, J. (1985) 'Parallel Currents in Natural and Social Sciences in Portugali, J. (ed.) Links Between the Natural and Social Sciences, Geoforum, 16, 227-238.

Portugali, J. (1986) 'Arab Labour in Tel Aviv: a preliminary study, International Journal of Urban and Regional Research, 10, 351-376.

Prigogine, I. (1981) <u>From Being to Becoming</u>, Freeman & Co., San Francisco.

Sandler, S. and Frisch, H. (1984) <u>Israel the Palestinians and the West Bank</u>, Lexington Books, Toronto.

Smith, A.D. (1971) <u>Theories of Nationalism</u>, Duckworth, London.

Smooha, S. (1978) <u>Israel: Pluralism and Conflict</u>, University of California Press, Berkeley.

Whyte, J. (1978) 'Interpretations of the Northern Ireland Problem: an appraisal' <u>The Economic and Social Review</u> 9 275-282.

Chapter 11

ETHNOREGIONAL SOCIETIES, 'DEVELOPED SOCIALISM' AND THE
SOVIET ETHNIC INTELLIGENTSIA

Graham Smith

According to Evans (1977, p. 414), the so-called era of
'developed socialism', a term coined by Leonid Brezhnev in
1967 to depict a new stage in the USSR's history, 'serves
both to convey a sense of progress and also to excuse the
failure of present Soviet institutions to match the standards of Full Communism'. In defending the introduction of
such an epoch, Soviet social theorists and policy-makers lay
particular stress on its goals of maintaining authoritative
political institutions and of securing economic progress
through the management of the economy as a single 'national
economic complex'. For the USSR's major non-Russian
ethnoregional societies, this implies two things. Firstly,
it implies, at least into the foreseeable future, a commitment to the peculiarly Soviet federal principle of maintaining state power in the ethnoregions with all this entails in
terms of highly centralised Party control administered
through its nationality-based political institutions in
tandem with a degree of cultural autonomy. For the state,
the continuation of such a territorial arrangement, which
for its architects was considered as a temporary phenomenon,
reflects necessity, not least because of the malleability of
ethnoregional feelings. This was shown in the defeat of the
abolitionist lobby during the period leading up to the
drafting of the 1977 Soviet constitution. Yet it is also
contended that in part, because of a federal structure which
provides for the 'flourishing' (ratsvet) of its ethnoregional societies, these peoples are made aware of the positive
contribution of the cultures of other nationalities to the
Soviet commonwealth of nations, and consciously identify
with them, a process, it is asserted, which strengthens the
unity of the Soviet people. Secondly, while recent policy
statements continue to reflect a commitment to further
improving the economic development and social modernisation
of particularly the more backward non-Russian Asian republics, the new ideology of 'developed socialism' argues that
Soviet society has progressed beyond needing to focus on

166

equalisation (vyravnivanie) of levels of socio-economic development among its ethnoregions to 'increasing the contribution of each of them to the consolidation of the country's overall economic complex to the economic power and defence capability of our multinational state' (resolution of the 27th Party Congress, Sovetskaya Rossiya, March 1986). The logic underlying such a focus is based on the claim that major ethnoregional inequalities have now been eradicated, and that the state is now in a position to concentrate on the rational specialisation of production in the common interests of the Soviet people while at the same time furthering ethnoregional convergence.

What therefore seems apparent is that as far as the Soviet leadership is concerned, the USSR has progressed sufficiently along the path to achieving regional socio-economic transformation to focus on All-Union concerns, yet progress has not been sufficient to warrant the abolition of the federal structure. Instead, it is contended that a new social and historical community of the Soviet people (Sovetskii narod) has come into being, a larger territorial community which is seen to have transcended nationality-based allegiances and is itself viewed as a significant contribution to the meaning of 'developed socialism'. While the removal of major socio-economic differences between ethnoregional societies is held to be a major contributor in the 'coming together' (sblizhenie) of the nationalities and an important prerequisite to the consolidation of Sovetskii narod, as Mikhael Gorbachev conceded at the 27th Party Congress in February 1986, such convergence is not in itself sufficient to erase nationality-based identities (Gorbachev, 1986, pp. 101-103).

Statements of this sort mark a departure from earlier expectations. As a specialist on the nationalities question in the Party journal of Soviet Georgia candidly states

> It was at the beginning of the 1960s that the voluntaristic approach to perspectives of the development of nations and national cultures and the attempt to run ahead first revealed itself in Soviet historical and philosophical literature. Alongside this, certain researchers "sincerely" believed that the 1960s generation would live in the period of communism, because those obsessed by the idea of "running ahead" with the construction of communism believed that it was only a matter of a few Five Year Plans. Consequently, in the opinion of certain scholars, only the national question remained to be settled - the "complete merging" (sliyanie) of nations and national cultures. These bad propagandists did not waste any time before announcing that the "complete merging of nations" was a task of the current moment. In order

> to speed up this "merging", they forgot that "a long road lay before the Soviet people on the way to communism - the road of socialist development and improvement - and displayed an irresponsible and frivolous attitude towards theory and the Programme's tenets." (Katcharava, 1986, pp. 2-3)

In contrast to the heady optimism of the early 1960s, today's leadership accepts that the speedy transformation of the economic base or social well-being of ethnoregional societies will not automatically erode people's consciousness of feeling ethnically distinct. The Party's recent and unprecedented call for a well thought through and scientifically substantiated nationalities policy reflects such pragmatism (Pravda, 22 December, 1982), as do policy speeches since the early 1970s which have avoided reference to the inevitable ending of such ethnoregional identities through the 'merger' (Sliyanie) of nationalities. Indeed one strand of Soviet scholarly research has moved away from the hitherto long accepted view that 'developed socialism' and an ethnoregional consciousness are necessarily incompatible. Bromlei (1983), for example, argues that ethnic identity (ethnos) is essentially a psychological phenomenon in which individuals see themselves as members of a tribe, nation, or nationality, none of which necessarily preempts a consciousness of being part of the Soviet people. It would therefore seem that with 'developed socialism', as Olcott (1985, pp. 106) puts it, 'at least as an interim goal, the Soviet leadership has introduced the idea of a united Soviet people in which political, economic, and social ideals are shared, but ethnic and cultural differences persist',

That the persistence of such ethnoregional identities constitutes a problem for the Soviet state is widely acknowledged by western experts of the nationalities question. Events such as language riots in Tbilisi following proposals to downgrade the status of the Georgian language (Parsons, 1982), protests by Lithuanians over the lack of religious and cultural facilities in their republic (Misiunas and Taagepera, 1983), and rioting by Kazakhs in Alma Ata following Moscow's dismissal of Kazakh First Party Secretary, Mukhamed Kunayev, and his replacement by a Russian (Pravda, 18 December, 1986), mark the most clearcut evidence of the malleability of nationalist feeling in a range of socio-economically and culturally variegated ethnoregional societies. Interpretations do however vary as to the communal strength of nationalist feeling, what processes may fuel those with a sense of nation-ness to take some form of political action, and the extent to which nationalism poses a threat to ethnoregional stability. First, there is what we can loosely label 'the modernisation thesis'. Here the argument goes that far from alleviating ethnoregional tensions and ensuring the structural assimilation of ethnoregional societies into a new supra-national, but Russian-dominated, value system, the universalising and secularising forces of modernisation have had the dysfunctional effect of further alienating native societies who resent their limited participation in a highly centralised state, and who are only too well aware of their disadvantaged status vis-à-vis the Russian core society. It is a thesis which emphasises the conflict-based nature of centre-ethnoregional relations and of the inability of the state to extricate itself from the unquestioned scope and strength of nationalist feelings (see, for example, Conquest, 1986).

An alternative school cautions against viewing state-ethnoregional relations simply in terms of a 'them' and 'us' conflict-type model. Here it is argued that the strength of an ethnoregional consciousness cannot be treated as an analytical 'given', that it necessarily exists as separate and can be easily disentangled from at least some form of identity with the Soviet state, and that we can automatically assume that blocked native upward mobility, differential regional development and limited participation in centralist political decision-making, or some other neat structuralist recipe, will necessarily lead to a heightening of nationality tensions, with the state heading for problems so great as to question even the very integrity of the presently-constituted national boundaries (e.g. McAuley, 1984; Rutland, 1984; Smith, 1985). For instance, McAuley (1984) has argued that the constitution of union republic-based ethnic societies, with their differing histories, cross-cutting social cleavages (e.g. class, party membership, urban-rural background, occupation), and sense of 'nation-ness', means that the conditions for the emergence of mass-scale support for a nationalist movement - which would pose a threat to stability - is not always in a form which would meet such a precondition. In a similar vein, Rutland (1984) warns against the tendency to treat 'the nation' as a socially homogeneous entity, as if all social groups share the same relationship in the cultural sphere. And finally, Smith (1987a) has argued that any analysis of state-ethnoregional relations must take on board the way in which the state draws upon such territorial mechanisms of control as the federal structure, passport system, and spatial language planning, as means of depoliticising ethnoregional societies by giving native elites in particular a vested interest in the socio-political status quo.

In this chapter I want to develop such a territorial dimension to state power and ethnoregional acquiescence by sketching out the rudiments of a theory of state-ethnoregional relations for 'developed socialism'. Attention will be paid to the role of one particular social grouping, the native intelligentsia. In order to highlight the differences between ethnoregional societies, I want to compare Latvia, one of the most socially advanced areas of the Soviet Union with a long history of national conscious-ness which dates back to the nineteenth century, and which culminated in a shortlived period of interwar independent statehood, with a developing society, Uzbekistan, whose people's sense of nation-ness is more recent, and where the Muslim religion is inextricably bound up with the Uzbek way of life. As part of the Soviet federal structure, both enjoy the same union republic status.

The State, Territoriality, and the Ethnoregions

In developing such a theory of state-ethnoregional relations it is important to begin by reminding ourselves that, as in all societies, the relationship between the state and its ethnoregions involves control over territory. While A. Smith (1986, p. 28) is right to note that 'territory is relevant to ethnicity ... because of an alleged and felt symbiosis between a certain piece of earth and "its" community', I think we must also bear in mind that subordi-nate nations and politically powerful state elites, to varying degrees, vie for physical control over territory and for allegiance from the communities that occupy such territory. The process or processes by which this is achieved can be referred to as territoriality, which Sack (1986, p. 19) defines as 'the attempt by an individual or group to affect, influence, or control people, phenomena, and relationships, by delimiting and asserting control over a geographic area'. That the Soviet state has an upper hand in such a relationship (to paraphrase Weber) stems from its successful claim over the monopoly of legitimate use of physical force within its territory. For the Soviet state, like any other multinational polity, power (which may be seen as the capacity to exert force effectively) and the le-gitimacy of the claim to use it are the very base on which control over its ethnoregions depends. Analysts who subscribe to a totalitarian view of state-society relations would emphasise the pivotal role that state coercion plays in maintaining compliance on the part of the ethnoregions. Yet as a number of commentators have noted (e.g. Bunce and Echols, 1980; Lane, 1984), since the early 1960s there have been conscious attempts by the state to change its strategy of gaining compliance by reducing reliance on coercion and strengthening consensual aspects. I will return to this later. What is important to stress is that such consent should not be envisaged as existing in a power vacuum. As Hoffman (1984, p. 139) reminds us, 'in societies where economic constraint is significant and political pressures pervasive, coercion and consent are unlikely to fall neatly into opposing camps'. In present-day Soviet society, force and consent are interrelated in complex ways. For instance, as a consequence of its territorial penetration into nearly every aspect of social, economic, and administrative life, the state possesses the facility to exercise force even if it does not choose to employ it. So even where force is not being used, it could be. As such, the essentially coercive state apparatus created under Stalin remains an important sanction at the disposal of the state and a basis to containing those centrifugal elements likely to challenge its power.

But just as we should not assume that it is simply the threat of force which ensures ethnoregional compliance, equally we should not fall into the other trap of assuming that individuals and social groups who do not revolt against the state, who do not initiate political opposition to it, give tacit consent to that political order, and thus accept obligation to it. The reasons why people obey are multifaceted and complex. These can include a sense of inevitability, a sense of being represented, of material interest in the socio-political status quo, of deference, fear, and resignation. For instance, during the 1960s and 1970s, Lapidus (1983, p. 189) has noted that a substantial congruence of elite and popular values is one important development

> even those Soviet citizens who reject many aspects of the system reveal great attachment to the dominant values of its political culture: order, discipline, paternalism, and social conservatism. They highly value security and stability in social and political life and appear to assign relatively low importance to political and civil liberties.

And even in those ethnoregional societies where nationalist feelings run high, an acceptance of the inevitability of Soviet rule has been detected. Dreifelds (1977,, p. 149) observes among Latvians that 'at present there is a certain unfocused anomie and alienation ... which seems to have loosened national feelings, pride, and resilience and which may impair group responses to further measures of denationalisation'.

One important aspect of manipulating ethnoregional acquiescence stems from the state's facility ·to allocate resources and distribute rewards in accordance with its own self-sustaining interests. That it is able to do so emanates from its near territorial monopoly over allocative and authoritative decision-making and from the inability of the ethnoregions, as part of an historical process, to control the growth of state centralisation. As a consequence, the state possesses a high degree of despotic power in which it can undertake a range of actions without routine, institutionalised negotiation with basic ethnoregional groupings. Yet at the same time, it must be seen to be democratic and participatory for it is on the basis of 'democratic centralism' that much of the state's claim to legitimacy in the ethnoregions rests. The existence of a federal structure and the ritual of elections among other aspects provide justification for Soviet power in the regions, although in actuality they provide little scope for localities to mould policy. However, in exchange for only the minimalist participation in the political decision-making process, benefits have accrued to ethnoregional societies in general, and to its elites in particular, as part of the package of 'developed socialism'. These include, most notably, a gradual improvement in regional living standards (particularly amongst late-modernising Central Asians), upward mobility, a degree of local control over cultural activities, and limited state interference in the 'second economy' (the latter particularly important to more 'inward looking' societies like Georgia and Uzbekistan). It is the basis of this uneasy compromise between territorial centralisation, on the one hand, and privileges of place, on the other, which I now want to explore.

Territorial Centralisation

Today's highly centralised state is a product of Stalinism which saw in territorial centralisation, rather than in the deconcentration of power, the most effective and rational means by which the socio-economic transformation of an economically backward, geopolitically vulnerable, and socially fragmented society, could be achieved. The early humanistic writings of Marx, which emphasised the dispersal of power and which viewed the state as a source of social alienation and political domination, irrespective of the dominant mode of production, had little pragmatic appeal to a post-Lenin leadership, who saw necessity in rapid industrialisation and in securing compliance against those elements of society considered as obstacles to achieving socialism. In contrast, such writings provided an important stimulus to anarchists, like the geographer Peter Kropotkin, who saw territorial decentralisation as fundamental to building socialism. Instead it was Marx's later and more technocratic model of the state, which seemed more logical and justifiable, in which change and securing a socialist society could only come about through the leadership of a strong and tightly centralised party-state machine, employing its coercive resources, whenever and wherever necessary, in order to achieve declared ends. As Harding (1984, p. 2) explains, selecting the strong and uncompromising state was in part a product of the lack of a dominant liberal tradition in Russia

> There never has been in Russia, and certainly not under the Soviet regime, any widespread conception of the individual or of social groups disposing of rights that can be exercised against the executive, openly to oppose its plans, limit its pretentions, or establish spheres into which it ought not to intrude.

Such a political tradition in combination with the practicalities of achieving 'socialism in one country', ensured the hierarchical subordination of the regions to the territorial centralisation of state power. The defence of local autonomy had no place in Stalin's Russia for it was contrary to state security and rapid industrialisation.

According to Bialer (1980, p. 141), the centralisation of state power presently 'is, if anything, greater than at the beginning of the Brezhnev era'. There was however some previous experimentation with decentralisation which has no doubt insured greater conservatism towards devolving power. It involved Krushchev's ill-fated attempt at giving managerial control to a number of regional economies through the setting up of regional economic councils (sovnarkhozy). The scheme not only led to problems of effectively coordinating the distribution of All-Union resources but to inspiring many ethnoregions, most notably Latvia, to put local interests before the state (Smith, 1979, 1982). Through the Party and its hierarchical regional structure, the post-Krushchev leadership continues to ensure the effective implementation of central policies without the regions challenging its power. Leading personnel in the union republics are appointed through the centrally-controlled party appointments procedure (nomenklatura system). Control from Moscow is also directly maintained through a central policy of earmarking particular key posts for non-natives, most notably, the post of Second Party Secretary whose functions include responsibility for the republic nomenklatura. Although it has been noted that the appointment of Slavs to such positions is nowadays based on managerial rather than on ideological skills (Miller, 1982), nevertheless, by appointing from outside and by keeping tenure to a minimum, the state forstalls the possibility that ties will develop with the local community, which functionaries like the Second Party Secretary are called upon to control.

Such strategies of territorial control do not however mean that the native union republic leadership is impotent and has no leeway in managing its region's affairs. Territorial politics does exist between the centre and the regions for nowadays, with a permitted pluralism, the union republic leadership bargains over a range of permissible issues, most notably resource allocation. This in itself is an important outlet through which to channel nationality feelings. However, as part of the central party apparatus, and having been given ethnoregional status and position by the centre, the native leadership is only too well aware that they are rewarded for managing the political status quo. Moreover, in contrast to the Krushchev period when territorial reorganisation and conflict were the order of the day, 'advanced socialism' has contained more significant elements of inter-elite cooperation and a search for consensus, a willingness if you like by the centre and its regional leaders to work towards common goals in which general interests are more clearly identifiable (e.g. planning for slow but purposeful overall economic growth, social welfare spending and its spatial allocation, more rational and less 'risk taking' planning, providing for more backward regions) (Bunce and Echols, 1980). In short, a 'bargain' of sorts has been struck between All-Union and regional interests. Part of the reason for the native leadership's willingness to work with the state must stem from Brezhnev's policy of putting greater 'trust in cadres'. Certainly party and state life has become less volatile and more comfortable under developed socialism. Analysis of key posts in the republic organisations show far lower rates of turnover during the 1970s than in the previous era (Blackwell, 1979, p. 33). Also, the frequency of 'outsiders' to appointments in key regional party leadership-positions has become less evident, a trend which Brezhnev did little to discourage and one which led to better promotion prospects for natives with the necessary specialist skills and local administrative experience (Moses, 1985).

Privileges of Place

The extent to which the state is seriously committed to eradicating major inequalities between its ethnoregions is the subject of much debate. While there is no doubt that at times in the USSR's history, particularly when powerful military and industrial interests can easily justify an increased share of the state budget, only lip-service has been to regional interests. However, it would be wrong to draw an analogy with classical colonialism in which the centre is seen to exploit its non-Russian regions. Rather, as Rykwin (1984, pp. 13-14) puts it, state policy can be considered as a type of 'welfare colonialism'.

It (the state) is willing to offer the inhabitants of these dependencies the equality of personal opportunities with the Russians, curtailing them only when

state security is in question. Ideologically inspired, it is willing to strive for economic, social
and educational equalisation between the ruling and
the ruled ethnic groups.

The extent to which the state is succeeding in
creating greater equality is also hotly debated, although
much of this debate has become increasingly entangled in
what criteria to use and of the comparability of such data
between regions and nationality groupings over time. Few,
however, would disagree with Jones and Grupp (1984) that
while major inequalities persist, a degree of convergence
did occur between the non-Russian and Russian nationalities
during the 1960s and 1970s. Such material improvement was
substantially greater in the cities and among the white
collar specialist class who acquired more and better
housing, enjoyed improved diets, and gained access to a
wider supply of goods and services, particularly consumer
durables.

Allied to such gains has been the expansion of
educational facilities and opportunities for native social
advancement, with the rapid growth of white collar and
technical occupations offering new employment opportunities
and enhanced social status. The process was especially
dramatic in the late modernising republics of Central Asia
and Moldavia, and has gone some way to rectifying the
spatial logic of Stalinist industrialisation which insured
that the growth of native specialists was uneven. If we
compare Latvian and Uzbek educational levels to the Russian
population then we find that in terms of student enrolment
in institutions of higher education, the 1960s saw considerable advance towards convergence, particularly amongst
Uzbeks. In more recent years, however, while student
enrolment in higher education continues to increase among
Latvians relative to Russians, for Uzbeks, probably due to
the state's inability to keep pace with their high rates of
demographic growth, the 1970s was a time when earlier gains
were reversed. The important point, however, is that the
native middle class became one of the fastest growing social
groups. They now constitute about a fifth of the Latvian
working population and about a ninth of Uzbeks.

Such native upward mobility owes much to the privileges of place which have accrued to the indigenous population as a consequence of union republic status. This
arrangement is a product of a Leninist ideology which viewed
a nation's autonomy as essentially territorial (in contrast
to the Austro-Marxists Bauer and Renner who saw language and
culture as defining a nation's autonomy) and of a pragmatic
political leadership, which saw in federalism a way of
compromising the nationalities, and so strengthening and
legitimising Soviet power. As a consequence, today, the

union-republic nationalities enjoy their own linguistic-based educational system, official status for the native language in public life, a flourishing national culture (albeit within a 'Soviet content') and an indigenisation (korenizatsiya) policy which discriminates in favour of natives in entrance to higher education, party membership, and top republican posts. Thus the preeminence of Russians in the central state apparatus is generally counterbalanced at the union republic level by the over-representation of natives, a process which has been especially positive in facilitating mobility in less developed societies like Uzbekistan with its smaller pool of qualified natives (Hodnett, 1978). Those nationalities further down the territorial hierarchy (autonomous republic status and below) receive less institutional supports. Consequently, for them, upward mobility often means adopting the Russian language and way of life, if not full cultural assimilation. Moreover, as non-Russian native institutional supports are largely confined to the arena of the union republic, its namesake population tends to remain rooted to the homeland where it can most effectively compete with incoming Russians for access to specialist jobs and positions.

The Native Intelligentsia

Before examining the role of the ethnic intelligentsia in such a territorial arrangement, a useful distinction can be drawn between two observable types of minority nationalisms, each with its own particular socio-territorial base. The first type is essentially rural-based but is also found in urban areas, particularly in the smaller towns. It is a nationalism connected, not so much with a coherent political ideology of anti-centralism or redistributive justice, but rather with a traditional way of life and concomitant system of values which neither industrialisation nor cultural assimilation have successfully eroded. It is inward-looking and localised, stripped of any coherent base or unifying ideology able to compete with Soviet statism. In heavily rural, polyglot societies like Uzbekistan it is a nationalism which is as likely to be based on local ethnic prejudices and age-old rivalries among the dominant Uzbek nationality (anti-Tadzkik, anti-Khorezum Turk, anti-Korean) as it is anti-Russian. It is also a nationalism largely untouched by competition with Russians for jobs or status in what remain comparatively socially homogeneous, rural, and non-Russian communities.

This contrasts with what Zaslavsky (1982) calls a 'nationalism of convenience'. This type is anchored in the urban-based intelligentsia or specialist class, the traditional bearers of nationalist ideas. It is among this stratum, particularly the humanistic intelligentsia, that we find the most outspoken elements of the native population, including, at one extreme, calls for a reconstituted and more decentralised polity, if not for the creation of a separatist state. Such demands are invariably bound up with the issue of human rights. As Kowalewski's (1980; p. 163) findings on the incidence of social protest show, 'the human rights movement - at least as regards protest demonstrations - remains a highly urban, middle-class, and adult phenomenon'. That the humanistic intelligentsia in particular should form such a source of opposition against centralism, party censorship, and other abridgements of human rights is in part due to the convergence of their ethnic and occupational interests and the function that they perform as transmitters of culture. Yet while their position, particularly their association with ensuring the reproduction of the native language and culture, singles them out as being especially problematic for the state, they should not be analytically divorced from the regional compromise already noted.

This is because the regionalisation of opportunities for native cultures provides an important basis for membership mobility and for preserving indigenous group attributes. As Hechter (1985, p. 22) notes in the case of ethnic regions, like Scotland and Spain's Basque country, 'Once a region has attained a degree of institutional autonomy by whatever means, this creates the potential basis for the development of a segmental cultural division of labour'. Native-based educational, administrative or legal systems, then, can furnish the conditions in which an ethnic middle class can flourish. Such a territorial arrangement in the Soviet case can help us to understand why, at the republic level, the native middle class has not needed to assimilate fully into the dominant (Russian) culture in order to gain position and status. The development of educational facilities in the native language, for example, which includes the opportunity to attend native-based schools for which the majority opt, has resulted in a demand for a large native intelligentsia equipped to read and speak the native tongue. Similarly, republican-based publishing institutions have brought into existence a very large grouping of native writers and journalists who quite naturally operate in terms of the culture with which they and their readers are familiar. The creation and continuation of native-based regional institutions have established niches for incumbents who adhere to the native culture and who are only too conscious of what they have gained as a result. They are also only too well aware that without the protective ringwall that such 'autonomous' institutions provide, they would no longer be able to enjoy the position and status in ethnoregional society that they have grown to expect.

It is rather paradoxical in such a production orientated society as the Soviet Union that we find in the ethnic regions a marked tendency of the middle class to cluster in the culturally related humanistic professions. According to data marshalled by the Soviet ethnographer, Kulichenko (1977, p. 97), for every 10,000 of their working populations, the Latvians and Uzbeks are particularly well represented in the following occupations (for comparison, Russians in parenthesis): teaching and the related humanistic professions, 586 and 641 respectively (640); artistic/creative, 66 and 20 (35); and administration, 321 and 181 (269). In other specialist jobs, natives are underrepresented with the notable exception of Latvians in economic production (1,092 per 10,000 compared with 354 for Uzbeks and 1,160 for Russians), which means that it would be difficult to sustain the argument that for geopolitical reasons, centrally-managed employment in the economic sphere has excluded an indigenous population able to effectively compete with incoming Russians.

That both native societies should predominate in the cultural sphere is certainly linked to the unwillingness and lack of need for incoming Russians to learn the indigenous language in regions where schools, the media and other institutions cater for their particular linguistic and cultural requirements (a fifth of such Russian communities know Latvian, and only one seventeenth Uzbek). Modernisation theorists are also right to argue that the clustering of Uzbeks in 'culturally-related professions' is due to a technical and higher educational system which demands of its technical recruits a fluency in Russian which a more rural-orientated and less developed Uzbek society does not possess. Equally, however, one should not underestimate the importance of cultural values in determining such occupational patterns. Indeed, Soviet policy-makers and economists have long since been concerned with trying to explain Uzbekistan's chronic shortage of local specialists in the production sphere which, they have often bemoaned, is related to culturally-specific traditional values, particularly the Muslim religion. As the late First Party Secretary of Uzbekistan, Sharif Rashidov, put it, 'Uzbek parents often feel that their children should work in science, culture, and the arts, and not in production' (Pravda Vostoka, 31 March, 1982). Many even prefer to work in the countryside in jobs in which they are overqualified, thus exacerbating city shortages. Lubin's (1984) detailed analysis of Uzbek society shows that natives often avoid particular occupations, most notably production-related employment, because of the low status that such posts hold in traditional Uzbek society. This contrasts with Latvians, where the traditional values of that society have judged production in a more positive way. Moreover, one should not underestimate the importance of the 'second economy' in such societies where jobs as in teaching carry many 'fringe benefits', so reducing the income gap between the humanistic and technical intelligentsia.

Throughout the 1960s and 1970s then, the state was able to provide means of support among the native intelligentsia by guaranteeing their employment within their namesake republic, by steadily improving their standards of living and, through the continuation of territorially-based native institutions and <u>indigenisation</u> policy, their upward mobility. Furthermore, there is also a passport system which regulates both geographic and social mobility into a region's largest cities, which furnishes the intelligentsia, concentrated as they are in these cities, with access to privileges which do not accrue to those ethnics living in smaller towns and in the countryside (Zaslavsky, 1982; Smith, 1987b). Providing such scope for the expansion, betterment, and place-specific employment of this class has borne an economic cost which a regime, in facing problems in arresting the economy's overall economic slowdown, may no longer be able to endure. This is especially the case with the more or less universal overproduction of a humanistic intelligentsia. Yet the uneven production of specialists more generally reveals a geography to this problem which does not augur well for either economic recovery or for harmonious nationality relations.

This can be seen by comparing Latvia with Uzbekistan. As in Georgia and Estonia, the Latvian economy is faced with overproduction of specialists which has already reached critical proportions. As the Soviet sociologists, Ostapenko and Susokolov (1984, p. 14) note,

> In these republics, even a worker with a speciality in short supply in the national economy will not readily find a job in keeping with his skill level. According to data from the Tallinn Bureau of Information and Labour Development, 90 per cent of workers applying for work in 1980 found work in their field during the course of the year, while 70 per cent of specialists did so.

In Latvia, such a trend is exacerbated not just by native upward mobility but also by migrants from other republics, primarily from the Russian republic. In the years 1979-84, for instance, just under two-thirds of the republic's population growth was attributable to migration, a demographic trend characteristic of most of the post-war period. This migration, although far from exclusively specialist, is overwhelmingly city-directed, and nowadays is as likely to be spontaneous as planned as Russians take advantage of the higher standards of living and western life-styles that the

Baltic offers.   Thus increased competition for scarcer positions, particularly where nationality differences are involved, will continue to fuel disquiet in the region.

The situation is different in Uzbekistan where there is an overall underproduction of native specialists.   High rates of indigenous population growth over the past two decades continue to outstrip available places in higher and technical education.   As a consequence, the region's specialist needs continue to be supplemented through migration, particularly from Russia, albeit at a lower rate than in the 1960s.   Mikheeva's (1975) comparative study of labour utilisation between Uzbekistan's towns and cities shows that a territorial cultural division of labour persists in which, with state investment primarily geared towards the larger cities, the inmigration of Slavic specialists and skilled workers has followed, with these settlements producing higher rates of employment and labour productivity compared with the overwhelmingly indigenous small towns.   Yet while the over-abundance of small-town rural labour could provide an important impetus to the region's development, barring massive capital investment in educational facilities, in the short term the region's shortage of specialist labour is unlikely to be resolved.

That the future economic and social progress of the republic may be in doubt stems from the USSR's need within a world economy to make itself more industrially competitive. This means further industrialisation, particularly in light manufacturing industry and in high technology.   Much of the credibility of the Gorbachev leadership depends on this. That such a strategy is likely to be at the cost of Central Asia and to the benefit of European republics like Latvia is a product of the spatial logic of economies of return and of a geography of skilled labour in which it is more expedient to utilise already available regional resources without stretching scarce capital investment.   Such an investment strategy would be welcomed by industrial and European interests, and could most certainly be justified by a Soviet leadership, on the grounds that major inter-regional inequalities have been resolved and that greater regional specialisation of production is in the interests of overall economic growth.   The prospect of Uzbekistan remaining essentially a supplier of agrarian commodities to European Russia, however, would carry no favour among the republic's middle class or for that matter its political leadership. Schemes to encourage northward migration of growing surplus Central Asian rural labour to meet Baltic/European Russian demand for unskilled manpower is also hardly likely to receive support, particularly if Uzbekistan has to continue to rely on an inflow of Russian specialists.

For the present, however, it is this increased competition between native specialists and incoming Russians for increasingly scarce positions which provides an important focus on indigenous concern throughout the non-Russian union republics. Of concern to this specialist class, and inextricably bound up with such competition, are recent central calls for linguistic conformity which for many is firmly seen as inextricably bound up with the promotion of Sovetskii narod and as part of a de-ethnicising process. Although the Russian language 'as a means of international communication' has always been viewed by the state as the linguistic cement which binds together such a polyglot society, at no time since the late 1950s has Moscow pushed its promotion with such vigour. Since 1978, the teaching of the Russian language has been expanded at all stages in the educational system, including the universities. For Latvians and Uzbeks, however, it is not so much the case of language shift to Russian which constitues a problem (there is no evidence to suggest from postwar census data that either nationality is abandoning its native tongue: at the 1979 census, 97.8 per cent of Latvians and 98.8 per cent of Uzbeks declared their nationality tongue as their first language); rather it is its spread, for language remains an important determinant of life chances. The more that specialists have to compete with Russians for jobs where the language of the workplace is Russian, or where Russian becomes a necessary precondition for access to higher education, the greater becomes the handicap of being Latvian or Uzbek, despite a knowledge of Russian.

The Search for Nation-ness

The native humanistic intelligentsia must also compete for the loyalty of other members of their nation with those elements of a standardising state which threatens to undermine the position of their nation and their people's well-being. As the transmitters of national culture and as its historicist educators, it is they who have been responsible for imbuing their populations with a geographic imagination of what is the nation, of defining its social boundaries,and of communicating the advantages of reconstituting relations between their territorial community and the centre. If the appeal is for separatism they must move away from an emphasis on primordial elements (e.g. ethnicity, language, distinctive cultural traditions, possibly also religion) to one based on identity with a territorial community in which social boundaries are firmly embedded and locked into a sense of national territory. How they go about such a task and the extent to which it proves an easy exercise cannot be taken as analytically given. Clearly it will be easier for some nations than others.

For Latvians, this sense of territoriality, of association between cultural boundaries and a territorial homeland, was established before incorporation into the Soviet state. Thus, 'Latvia' has political and cultural meaning as well as being an administrative fact of life. The years of independent statehood, particularly the 1920s when its peoples can claim to have established a liberal democracy, provides the community with a benchmark for judging contemporary reality. Latvia has also one of the highest proportions of native intelligentsia, which was well developed before the Soviet period, and an economy far in advance of most other ethnic regions. Rather than seeing its industrialisation and high living standards as a consequence of Soviet rule, one finds some evidence of even those managers in economic production putting emphasis on such a transformation as being Latvia's achievement (Shryock, 1977). For historical reasons, one also finds Latvians 'looking outward' for economic comparisons with Scandinavia and with Eastern Europe. Moreover, there is much general resentment at the pace of this imposed industrialisation: besides relegating Latvia's traditional and successful rural economy, centrally-directed industrialisation is also seen as responsible for the republic's ethnic metamorphosis (Latvians now barely constitute a majority in their republic), for exacerbating housing shortages, and for environmental problems, all of which have been seized upon by dissidents, and occasionally receive an airing by loyal middle-class dissenters (Smith, 1982).

All these conditions make Latvia politically one of the most problematic for the Soviet state. Yet one should not underestimate the state's ability to control through territorial and other means the development of a nationalist movement. Despite empathy with the Poles about economic mismanagement and the insensitivity of bureaucratic centralism, the political protests in 1980-1 in Poland which gave rise to the emergence of Solidarity were easily contained in the republic, partly because of a loyal, local political leadership fulfilling its function as the centre's manager of ethnoregional problems. No doubt fearing geopolitical spillover effects, it attempted to neutralise the situation by employing punitive measures (e.g. the rounding up of dissidents) in combination with trying to redress some of the community's grievances. Most notable in this regard was First Party secretary, Avgust Voss, who took the opportunity at a major conference held in Latvia in 1982 on 'nationality relations under developed socialism' to call for a more balanced approach in language policy and for the development of a genuinely more bilingual policy (Sovetskaya Latviya, 29th June, 1982). No doubt the latter comment was also aimed at the local Russian community's linguistic intransigence.

Such ethnoregional anxieties and dissatisfactions are not simply acted out in costumes borrowed from past generations. Where national identities are not novel constructs - as in the case of Latvia - they invariably entail the transmutation of older categories to meet new needs in new ways. One such response which can be identified is a new sense of social realism as Latvians respond chameleon-like to the new circumstances in which they find themselves. Rather than viewing separatism either as feasible or for some sub-groups as even desirable, it involves focusing on what Allworth (1977, p. 16) calls 'a suitable group maintenance system'. By this he means that rather than focusing on those institutions and ways of life now missing (e.g. national statehood, economic independence, religious and cultural self-determination, and the like), a feature of nationality nostalgia during the late 1940s and 1950s, the 'supports' that are increasingly emphasised include greater accent on literary and artistic expression, a more lively appreciation for preserving local values in the countryside, and an increased emphasis on the subtleness and richness of the native language. Such a reconstituted sense of national awareness has also been commented upon by a number of Soviet ethnographers (e.g. Drobizheva, 1985).

In contrast to Latvia, Uzbekistan is not the creation of a nationalist consciousness but rather of a Soviet state, which in dividing up Muslim Central Asia/Turkestan into union republics, gerrymandered the consolidation of a weakly developed Uzbek linguistic and territorial base. The development of an Uzbek national consciousness is therefore very much a result of Soviet rule. In the Central Asian case, Moscow has also used territory to obscure nationality, for Uzbekistan is not simply the homeland of Uzbeks but of numerous indigenous nationalities. Consequently, a newly established Uzbek intelligentsia has sought to create an Uzbek identity out of a previously non-existent one. The task of establishing an Uzbek lineage before the Soviet period is not that easy, as a recent article by the Uzbek historian, Polatov (1984), highlights. In arguing that Uzbeks only reached nationhood in the post-revolutionary period, he suggests that for a full awareness of themselves as a socialist nation, it is nonetheless important to establish a history and links stretching back into antiquity. This is a project, he contends, which is not incompatible with the development of Sovetskii narod. As a number of Uzbek historians and ethnographers have pointed out, theirs is an ancient history which can be traced back to the glories of Samarkand and Bukhara and which, as some would have it, is, at least spiritually, their cultural debt, and influenced by traditional Islam rather than the role of Russia as 'elder brother'. Yet the difficulty of disentangling such an identity with Turkestan or the Muslim

religion is no easy task. As a secular ideology, nationalism has to compete with the Islamic umma if it is not to
aspire to anything short of territorial separatism. Gorbachev's recent speech in Uzbekistan on the need to launch a
concerted attack on the Muslim religion, no doubt spurred on
by fears of Islamic fundamentalism in South-West Asia
spreading to the Soviet Union, does make Uzbeks even more
aware that their Muslim way of life is not compatible with
aesthetic state standardisation (Pravda Vostoka, 25 November, 1986). It does not, however, put being 'Uzbek' at the
top of a self-defining political agenda.

Uzbek economic development is also a consequence of
Soviet rule as are its intelligentsia and ruling elite. That
their region does not compare well with other parts of the
USSR may be of less critical relevance than the fact that in
comparison with Muslim countries on their southern borders,
standards of living are generally higher. The state makes a
great play of the fact, using Muslim Central Asia (and in
particular its mullahs as ambassadors) as the economic model
for other Muslim countries to emulate. Moreover, elites have
stability on their side; the Red Queen does not come along
every so often and knock off their heads. Also one should
not underestimate the way in which the republic's rurality
and the failure of the state to penetrate parts of its
cultural economic life plays in pacifying political dissent.
In particular, the private sector is important. The private
plot (an important aspect of the second economy and on which
Gorbachev has committed himself to retaining) is strong;
there is also more privately owned housing in Uzbekistan
than in most other areas. The 'second economy' also extends
into the public sector (e.g. kiosks selling privately
processed goods, factories working 'on the side' in terms of
their regular production). While no regional society is
immune from such practices, it is more widespread in
Uzbekistan than in Latvia. As Rykwin (1984, pp. 5-6)
explains, 'This is made possible by the very nature of
Central Asian society: family unity, extended family ties,
local friendships (mestnichestvo), ethnic-religious solidarity, all combine to shield private dealings from the public
(or, rather, governmental) eye'. In contrast, Latvians lack
the shield of extended family (or tribal) ties which in
Uzbek society is difficult to penetrate even by an outsider
with the same ethnic background.

Conclusions

I have argued that the ethnic minority regions have in part
been depoliticised by privileges of place. In combination
with territorial centralisation and the consequences this
has in providing a strong power base at the centre, the

state has been able to provide, during the period of advanced socialism, ethnoregional stability. Pacifying in particular the humanistic intelligentsia has been at considerable economic cost. Whether or not Gorbachev's USSR will be able to continue to provide a balance between their material and decentralising interests, on the one hand, and economic recovery and the continuation of state centralisation, on the other hand, is something that we can only guess.

Acknowledgements

I am grateful to Robert Parsons for showing me his translation from Georgian of the Katcharava (1986) article.

References

Allworth, E. (ed.) (1977) Nationality Group Survival in Multi-Ethnic States: Shifting Support Patterns in the Soviet Baltic Region. Praeger, New York.

Bialer, S. (1980) Stalin's Successors: Leadership, Stability and Change in the Soviet Union. Cambridge University Press, Cambridge.

Blackwell, R. (1979) 'Soviet Cadres Policy', Problems of Communism 28, pp. 29-42.

Bromlei, Yu V. (1983) 'Ethnograficheskoe Izuchenie Sovremennykh Natsional nykh Protsessov v SSSR (k 50-letiyu ordena druzhiby Narodov Instituta Ethnografii AN SSSR)', Sovetskaya Ethnografiya, 2, pp. 4-13.

Bunce, V. and Echols, J. (1980) 'Soviet Politics in the Brezhnev Era: "Pluralism" or "Corporatism?"' in D.R. Kelley (ed.) Soviet Politics in the Brezhnev Era. Praeger, New York, pp. 1-26.

Conquest, R. (ed.) (1986) The Last Empire: Nationality and the Soviet Future. Hoover Institution Press, Stanford.

Dreifelds, Y. (1977) 'Latvian National Demands and Group Consciousness since 1959' in G. Simmonds (ed.) Nationalism in the USSR and Eastern Europe in the Era of Brezhnev and Kosygin. Detroit University Press, Detroit, pp. 136-56.

Drobizheva, L.M. (1985) 'Natsional' noe samosozanie: baza Formirovaniya i sotsial'no - Kulturnye Stimuly Rezvitiya', Sovetskaya Ethnografiya, pp. 5-25.

Evans, A.B. (1977) 'Developed Socialism in Soviet Ideology', Soviet Studies, 29, pp. 409-28.

Gorbachev, M.C. (1986) Politicheskii doklad tsentralnogo Komiteta Tsentralnogo Komitera KPSS XXXVII S'ezdy

Kommunisticheskoi Partii Sovetskogo Soyoza, Politiz-
dat, Moscow.

Harding, N. (1984) Socialism, Society and the Organic
Labour State, in N. Harding (ed.) The State in
Socialist Society. Macmillan, London, pp. 1-50.

Hechter, M. (1985) 'Internal Colonialism Revisited', in E.
Tiryakian and R. Rogowski (eds.) New Nationalisms of
the Developed West. Allen and Unwin, London, pp.
17-26.

Hodnett, G. (1978) Leadership in the Soviet National
Republics. A Quantitative Study of Recruitment.
Mosaic Press, Oakville).

Hoffman, J. (1984) 'The Coercion/Consent Analysis of the
State under Socialism', in N. Harding (ed.) The State
in Socialist Society. Macmillan, London, pp. 129-149.

Jones, E., Grupp, F. (1984) 'Modernisation and Ethnic
Equalisation in the USSR', Soviet Studies, 36, pp.
159-184.

Katcharava, I. (1986) 'Erebis Aqvaveba da Urt'iert'gamdid-
reba Sotsializmis Monapovaria', Komunisti, 21 February, pp. 2-3.

Kowalewski, D. (1980) 'Trends in the Human Rights Movement',
in D.R. Kelley (ed.) Soviet Politics in the Brezhnev
Era. Praeger, New York, pp. 150-81.

Kulichenko, M.I. et al. (1977) Natsional'nye Otnoshenia v
Rasvitom Sotsialisticheskom Obshehestve. Mysl',
Moscow.

Lane, C. (1984) 'Legitimacy and Power in the Soviet Union
through Socialist Ritual', British Journal of Political Science, 14, pp. 207-17.

Lapidus, G. (1983) 'Social Trends' in R. Byrnes (ed.) After
Brezhnev. Sources of Soviet Conduct in the 1980s.
Frances Pinter, London, pp. 186-249.

Litvinova, G., Urlanis, B. (1982) 'Demograficheskaya
Politika v SSSR', Sovetskoe Gosudartstvo i Pravo, 3,
pp. 38-46.

Lubin, N. (1984) Labour and Nationality in Soviet Central
Asia. An Uneasy Compromise. Macmillan, London.

McAuley, M. (1984). 'Nationalism and the Soviet Multiethnic
State', in N. Harding (ed.) The State in Socialist
Society. Macmillan, London, pp. 179-210.

Mikheeva, V. (1975) 'Trudovye Resursy Malykh i Srednikh
Gorodov Uzbekistana i Perspektivy ikh Ispol'zovaniya,
Unpublished doctoral thesis, Tashkent.

Miller, J.H. (1982) 'The Communist Party: Trends and
Problems' in A. Brown and M. Kaser (eds.) Soviet
Policy for the 1980s. Macmillan, London, pp. 1-34.

Misiunas, R.J. and Taagepera, R. (1983) The Baltic States.
Years of Dependence, 1940-1980. C. Hurst, London.

Moses, J.C. (1985) 'Regionalism in Soviet Politics: Continuity as a Source of Change, 1953-1982', Soviet Studies, 37, pp. 184-211.

Olcott, M.B. (1985) 'Yuri Andropov and the "National Question"', Soviet Studies, 37, pp. 103-17.

Ostapenko, L.V., Susokolov, A.A. (1984) 'Etnostsial'nye Osobennosti i Vospriozvodstva Intelligentsii', Sotsiologicheskie Issledovaniya, pp. 10-16.

Parsons, R. (1982) 'National Integration in Soviet Georgia', Soviet Studies, 34, pp. 547-569.

Polatov, H. (1984) 'Tarkihiy ong: Milliy va International Jihatlar', Sovet Ozbekistoni, January, pp. 2-3.

Pravda, (1982) 22 December

Pravda, (1986) 18 December.

Pravda Vostoka (1982) 9 January.

Pravda Vostoka (1982) 31 March.

Pravda Vostoka (1986) 25 November.

Rashidov, S.R. (1978) 'National'nye Otnosheniya v usloviakh Rasvitogo Sotsialzma', Voprosy Filosofii, 19, pp. 3-18.

Rutland, P. (1984) 'The "Nationality Problem" and the Soviet State' in N. Harding (ed.) The State in Socialist Society. Macmillan, London, pp. 150-178.

Rykwin, M. (1984) 'National Symbiosis: Vitality, Religion, Identity, Allegiance', in V. Ro'i (ed.) The USSR and the Muslim World. Allen and Unwin, London, pp. 3-15.

Sack, R. (1986) Human Territoriality. Its Theory and History. Cambridge University Press, Cambridge.

Schroeder, G.E. (1986) 'Social and Economic Aspects of the Nationality Problem' in R. Conquest (ed.) The Last Empire. Nationality and the Soviet Future. Hoover Institution Press, Stanford, pp. 290-313.

Shryock, R. (1977) 'Indigenous Economic Managers', in E. Allworth (ed.) Nationality Group Survival in Multiethnic States. Shifting Support Patterns in the Baltic Region. Praeger, New York, pp. 83-122.

Smith, A.D. (1986) The Ethnic Origins of Nations. Blackwell, Oxford.

Smith, G.E. (1979) 'The Impact of Modernisation on the Latvian Soviet Republic' Co-Existence, 16, pp. 45-64.

Smith, G.E. (1982) 'Die Probleme des Nationalismus in den Drei Baltischen Sowetrepubliken Estland, Lettland und Litauen', Acta Baltica, 2, pp. 143-77.

Smith, G.E. (1985) 'Ethnic Nationalism in the Soviet Union: Territory, Cleavage, and Control, Environment and Planning C. Government and Policy, 13, pp. 49-73.

Smith, G.E. (1987a) 'Administering Ethnoregional Stability: the Soviet State and the Nationalities Problem', in C. Williams and E. Kofman (eds.) Community, Conflict, Partition and Nationalism. Croom Helm, London.

Smith, G.E. (1987b) 'Privilege and Place in Soviet Society', in D. Gregory and R. Walford Horizons in Human Geography. Macmillan, London.

Sovetskaya Latviya, (1982) 29 June.

Sovetskaya Rossiya (1986) 7 March

Zaslavsky, V. (1982) The Neo-Stalinist State. M.E. Sharpe, N.Y.

Chapter 12

THE OCCURRENCE OF SUCCESSFUL AND UNSUCCESSFUL NATIONALISMS

H. van der Wusten

The story of nationalism in Ireland has often been told in
the following manner. The Irish having an agelong separate
identity gradually reawaken after prolonged suffering under
the English yoke. In a number of ever more impressive shows
of strength they acquire national independence for the major
part of the country with the last instalment just around the
corner. As Parnell said when the process was halfway: 'No
man has the right to set a boundary to the onward march of a
nation'. Quite so. An alternative is possible. It
narrates a number of discrete events, all different efforts
to get more political autonomy varying as to aims, means and
supporters, each time triggered by contemporary circumstances and always having a different course on account of
the intricate process of moves by the Irish and countermoves
by a strong adversary. In the encounter at the end of World
War I some successes were gained, and partition then had to
be accepted. Would there be a next round? In the end there
was something of the sort triggered by new circumstances
(van der Wusten, 1977).
        The first version fits the needs of the nationalist
movement itself and of those who look for structural
explanations. The second version gets a sympathetic hearing
from the opponents of that particular nationalism and from
those who think of history largely in terms of an innumera-
ble series of contingencies. There is no doubt that both
versions are wrong. The first overstates continuity, the
second one wrongly negates it. In order to improve on the
first version we have to be more serious about the interrup-
tions of the march of the nation. In order to improve on
the second one needs to ask why it is that some collectivities have so much more often marched as nations than
others. In combination they elicit the following questions:
what are the appropriate units of analysis in the study of
nationalism? What are the conditions for nationalism to
occur? What is the yardstick for success (thus failure)?
How to explain success/failure?

It is appropriate to phrase the questions in this way in spite of the obvious difficulties in answering them. There is a certain attraction in looking at final results or at the state of a phenomenon at the end of a period and then moving backward in time either in search of explanations or considering concurring phenomena that may be interpreted as such. In this way occurrences cannot properly be explained because non-occurrences are neglected and there is a tendency not to study failures on a par with successes. This is the same point Tilly (1975) has stressed in his critique of the studies on state-formation and political development. In this way teleological reasoning becomes part and parcel of the research design. It is evident, however, that the alternative I have put forward engenders its own difficulties. The area under discussion is extremely broad and I will not be able to put more or less theoretically grounded suggestions to a rigorous empirical test. I will only bring forward some tentative ideas and illustrate these by way of examples. This obviously introduces other sources of bias.

Nationalism(s)

Nationalism can be described as a doctrine proposing some measure of political autonomy for and in the name of a social collectivity assumed to be homogeneous and cohesive. This collectivity is the nation. In order to become socially relevant the doctrine of nationalism needs a carrier. This is the nationalist movement. The aims propounded by the doctrine of nationalism and its movement are the achievement of some measure of political autonomy (we call this state-making after the highest level of autonomy available) and/or the realization of that homogeneous and cohesive collectivity supposed to be its founding stone (nation-building).

Nationalism has become an ever more widespread notion since the days of the American and French revolutions. Social movements and singular acts have occurred under its aegis. Nationalism has been in operation in a huge diversity of environments. All these nationalisms have thus been tinged by diverse circumstances. They may be classified according to at least three dimensions: the nature of nationalist aims; the ways in which movements operate; and further ideological content of the doctrine.

Nationalist aims differ according to the degree of political autonomy they seek. This may vary from forms of partial internal self-government to full sovereignty. Although in each and every nationalism the ideas of political autonomy and that of a nation are to be found, they can be pursued with different priorities. State-making or more moderate forms of self-government may get priority over nation-building or the other way round. In most cases this is demonstrated more clearly at the level of the actual operation of the social movement than at the doctrinal level proper. There may in many movements be two inclinations at the same time: one towards state-making or less extreme versions of political autonomy probably headed by political professionals and those who aspire to these roles; and one towards nation-building headed by professionals in occupations with a socializing function like teachers. It depends on the commanding faction which way the movement goes. Shifts in the orientations of persons occur (e.g. Patrick Pearse's career in the Irish case). If shifts in emphasis occur, continuity of persons involved and continuity as regards population and territory where state and nation are thought to be situated have to provide the basis of judgements on continuity.

Movements operate as a rule by leaps, campaign-like. After periods of apparent standstill they may change into a mode of full operation before returning to their initial slumber, sometimes quite suddenly. The question then arises, what is a movement and what happens during these outbursts of action? A movement may be considered a set of social units oriented towards some common aim for which the units are prepared to pool their resources and use them collectively, plus a leading unit whose job it is to signal that pooling of resources should start and that subsequently controls their collective use. In other words there is a voluntary element and there is an element of hierarchy and dominance. Once a campaign has got underway, dominance takes precedence over the voluntary element but the leading unit is normally not able to maintain control and direct the use of resources indefinitely and willingly or unwillingly the campaign comes to an end. The voluntary element takes precedence again and the movement largely changes into a state of potentiality.

It will be evident that campaigns may come to nothing for a number of reasons. Orientations of units and their acceptance of the leading unit may have been wrongly estimated. Errors may have been made in the assessment of available resources. The leading unit may not be able to communicate effectively with its followers. Circumstances may change in unanticipated ways through the interactive nature of the conflict in which the movement is involved etc. etc.

Looking for the appropriate unit of analysis there is a major problem with regard to the time dimension. Because a social movement generally, thus a nationalist movement, may for prolonged periods of time be in a state of potentiality and new developments will occur, the life or death of that particular brand of nationalism will only be evident once a campaign gets started. It may turn out that nationalism has quietly died or has merged with elements from other doctrines, but it may also come to life again quickly once called upon (a short military sortie or a royal wedding may suffice). The question is, how can we know that there is a potentiality of nationalism during these dead periods? In fact we cannot, at least not very exactly. The only thing we can do is recognize that if at two different moments in time two sets of social units which have some sort of meaningful continuity (region etc.) burst into nationalist activity, there is a definite probability (no certainty) that the nationalist doctrine was carried in the meantime by a social movement in a state of potentiality. It was not acted upon for the time being. We should nevertheless look for signs (common symbols, mutual orientations and the like) and then conclude about the continuous or discontinuous nature of nationalism in this instance. Another more obvious problem with regard to the unit of analysis in the study of nationalism is the precise extension of the nation and the territorial claims this entails. These may be hazy along the edges.

Nationalist movements may be broadly based and popular with political formations like a congress-party or a mass-party with a central bureaucracy, or they may be organized on a restrictive basis following elitist principles like a secret society or a vanguard party. And we should expect all sorts of mixed types. The major Irish campaigns by no means cover the whole spectrum but they demonstrate interesting differences in this regard. O'Connell's campaign in the 1840s was very broadly based but at the same time the steep social hierarchy was totally respected within the movement. Although popular it was by no means egalitarian. Parnell's campaign was also mass-based and at the same time far more egalitarian. There was less respect paid to the socially privileged. An exception may have been the leader himself, until his fall in 1890/1. The 1919-1921 campaign aroused popular feeling on an egalitarian basis. It was a mass-campaign, but at the same time a secret society (the IRB), or at least some of its members purporting to have a privileged point of view with regard to the course of the future tried to stay in the lead. This was one important source of friction among the Irish leadership during these years. The IRB had earlier launched an abortive campaign in the 1860s which is a typical example of an elitist movement driven by a secret society.

Nationalist movements have various <u>repertoires of social action</u> at their disposal, an important differentiating characteristic being the role of violence for its intrinsically destructive capacity and its significance in polarizing and intensifying conflicts. The role of violence is difficult to assess unambiguously. For the major Irish campaigns, for example, we may say that O'Connell's was bloodless. Parnell's campaign made use of the boycott (the term was coined in the preceding Land War). Violence was officially renounced but in fact tolerated to bring the potential following in line and deter opponents. During 1919-1921 two subcampaigns were conducted at the same time, one violent and one largely non-violent, with coordination and overlap but also with internal contradiction and frictions.

Secret societies and vanguard parties are generally inclined to violence if they are not very much an accepted elite group in society already. This is so because they tend to perceive themselves as in a hurry which encourages the use of means thought to be extremely effective. They also have a lack of other options if they want to realize their goals without a broadly based support. In Hirschman's (1970) terms their voice is violence, otherwise they have to accept exit or loyalty. These two last courses may in fact have been followed quite often without leaving a trace. This is quite improbable for mass movements. Once they have reached a certain size they cannot so easily be blotted out from history. Depending on circumstances perhaps more than on principles, mass movements may be more or less violent.

Occurrence

What do we mean when we state that a specific nationalism occurs, is present? There should be some nationalist doctrine accepted by a social movement, that is potentially active or in full action, and considers this to be the first item on its agenda. The movement may still be extremely small compared to the size of its claimed nation. But in order to be socially relevant it should be seen as relevant by others, as an adversary or an ally. As these perceptions will be widely different in space and time, a threshold value beyond which nationalism definitely exists is difficult to operationalize. This is something that more often happens in the social sciences (think of a concept like integration), but it is nevertheless an important problem because in this way we lack a sound baseline from where to calculate rates of success/failure.

What is needed for nationalism to occur? Apparently a version of the doctrine, a leading unit that fosters and propagates it, some following and by implication a relation between the two. The conditions for the appearance of these elements should be stated.

The leading unit contains people who have internalized the nationalist doctrine and are inclined and able to act accordingly. They may be ideological activists drawn from the intelligentsia or the heads of state bureaucracies who consider nationalism the self-evident starting point for taking decisions. Since nationalism became a doctrine at the end of the 18th century such people have always been around in many places in Europe and North America. The doctrine became generally known by living example. And as this type of people became part of the social structure in more and more countries, nationalism was spread. The presence of such people is probably a necessary but not a sufficient condition for nationalism to occur. It is a necessary condition because of the particular sensitivity of these categories to forge links between general ideas and practical behaviour. But it is not a sufficient condition because they have to choose for this particular doctrine.

In order to have a persuasive case for a nationalist doctrine, an image of the nation is necessary. As some of the funny stories in the 'Invention of tradition' (Hobsbawm and Ranger, 1984) suggest, the only thing needed might be somebody with a lively phantasy but this is apparently not the only condition to be fulfilled. The nation as an image must have some roots in historical reality. Once upon a time a cohesive and homogeneous unit was there and a tangible link with the present can be constructed, be it (the remnants of) language, religion, custom, institutions or the like.

For the leading unit to become constituted and to act upon the nationalist doctrine, there must be a sense of urgency. Reality does not live up to expectations based on the image of the nation and the political autonomy that is its of right. Either the real nation is felt to be threatened or deteriorating or political autonomy is threatened or inadequate. This may happen under two sets of circumstances.

Under the first the actual situation is thought to worsen. This is the case when a system of national education is thought to uproot the native language and way of thinking: The Murder Machine, the Irishman Pearse called it. But this is just the same as those who fear the undermining of a nation that is congruent with a state, from granting a right to use the minority language in broadcasting and the like. Only the preferences differ. The same applies to the threat of political autonomy. There may be external or internal threats to the integrity of the state which then kindles nationalism at the scale of the state (Smith, 1981). But the withdrawal of rights of autonomy within a state may also kindle a nationalism, only another one. In the Irish case agreed home rule legislation was shelved in 1914 by the British government. This threatened expectations of impending political autonomy which was one of the circumstances that facilitated the 1919-1921 campaign.

There is another set of circumstances that may produce the same results. If somewhere else, in a country where circumstances are pretty well comparable to those we are looking at, more political autonomy is granted or the image of a nation materializes to an extent, the present situation of our case is seen as inadequate and there is an urge to imitate. In addition successful role models are then available and learning can be quick. Decolonization, once underway, is a very good example and a feeling of deprivation plus the urge to imitate former colonies have also been brought forward as possible explanations of the sudden revival of so-called ethnic nationalism in Europe.

Irish nationalist movements leaned heavily on others in a doctrinal sense (finance, particularly for the later campaigns, came primarily from the US). People who moved in O'Connell's circle had fought in Latin America against the Spanish. The Young Ireland faction in his movement was directly using the example of Young Italy. The original organizer of the IRB, James Stephens, got his education as a revolutionary in Paris in the 1850s. The Home Rule movement that was eventually headed by Parnell, started in the early 1870s after Canada had obtained internal self-government. It was directly inspired by this example. The Irish insurrectionaries of 1919-1921 reacted to some extent to the American President Wilson's rhetoric on self-determination and one of the earliest small-scale starts of this campaign can be traced to the impression the Boer War around 1900 left in Ireland, where the fight by the Boers was seen as an example to be followed.

Williams (1986) has recently proposed that these two mechanisms worked side by side as the teachings of the French revolution spread over Europe. On the one hand the fact of the French revolution happening was an object lesson for people all over Europe to start their own nationalism (deprivation plus imitation). On the other hand the conquest of other countries by the French armies threatened the respective images of the nation in these other countries. Leading units became aware of and could then capitalize on this widespread feeling of actual threat among their potential following as well. In France itself the fact of war forged nationalism also. A third mechanism might be that the French not only showed the way by their demonstration in France but also taught some practical lessons in nationalist rhetoric while occupying their neighbours, in this way also accelerating the propagation of this innovation wave. This example also emphasizes again that the difference in attitudes between leading units and others may in fact be quite relative, before as well as after the nationalist movement has been swung in campaign-like action. But if during quiet periods when the movement is demobilized no leading unit has been formed or has maintained itself, no action of any duration or effectiveness is possible.

Those who follow the leading unit, necessary for nationalism to come into existence, have two characteristics. They are aware of their link to the imagined historic nation. That is, they foster the common heritage and use the required vocabulary, or even a separate language. Secondly they accept the leading unit as such. This is in both cases largely a question of learning behaviour facilitated by the tangibility of the elements to be learned, the efforts of the leading unit using the positive and negative sanctions at its disposal and the effects of the social environment, or society at large.

Society at large plays a crucial role in the presence or absence of nationalism and in the kind(s) of nationalism generated. First of all, for movements at some scale to exist one needs sufficient infrastructural and personal capabilities to get mobilization off the ground. Deutsch's (1961) concept of social mobilization refers to this. It is not so much directed to the actual rate of mobilization as it is to the capacity to bring it about. It means that in traditional society the chances for any movement, such as nationalism, are smaller.

The second general point in this regard is concerned with strategic options and chances. Who are the other parties in the field where nationalism should come off the ground that will either facilitate it or limit its chances?

Three general situations can be distinguished. The first is where nationalism is the only doctrine around and it has simply to conquer that part of the population with the capacity to become mobilized. This seems a rather rare situation. Of course such nationalism is further facilitated as the territory that should be its nation's is threatened from the outside.

The second situation has one nationalism aiming at a population and territory at large and one or more smaller nationalisms inside. These are the so-called ethnic nationalisms in Europe pitted against their respective states but also Arab nationalism versus Egyptian, Syrian, Iraqi, Algerian, etc. nationalism. To look at ethnic nationalism as different in kind from state nationalism may, I think, be confusing. Both nationalisms are driven by the same forces. They are in opposition to the extent that ethnic nationalism cannot be satisfied with some measure of autonomy within a state. If it cannot, there is a zero-sum game.

A third case is where nationalism competes with well-articulated alternative doctrines and movements like those of farmers, christian-democrats, communists and the like. These will in most cases be oriented to the state but they may rest upon a largely regionally concentrated following. Particularly in this last case some confusion with nationalism may occur and there may be conflict about which issue should come first. In the various stages of

Irish nationalism this is quite clear (there is always the triangle of farmers, Catholics and Irish national interests to cater for). The specifically nationalist appeal can at certain moments be aroused for a variety of reasons: urgency compared to other issues at the regional level, political tactics, the nation at large is in peril and will not be able to strike back and the like. If there is a number of well-established alternative movements without such regional concentration, only an overriding sense of urgency (attack, occupation) may be able to bring nationalism back to the fore.

Success

Once nationalism has got underway what will happen? For a start we will be interested to know to what extent nationalism gets realized and how much time is needed to bring this about. However clear this appears to be, it is a highly ambiguous statement because it leaves open whether we will assess nationalism's successes in terms of the aims of each particular doctrine or in terms of nationalism generally. Some nationalist doctrines aim at limited forms of self-government only while for others the goal is full sovereignty. If we take full sovereignty as a universal yardstick doctrines aiming at limited self-government will for that reason alone very probably be less successful as a general rule. But are we well advised to take such a course?

The principle of self-determination and the actual process of decolonization have strongly impressed the fullest political autonomy thinkable as the self-evident yardstick of the realization of nationalism. This may however be the dominant opinion of a limited historical period only, with other periods far less doctrinal in terms of acceptable solutions to a longing for political autonomy. The Irish case again provides impressive evidence of this variability. Full sovereignty only became the dominant creed of Irish nationalism when Sinn Fein defeated the Irish nationalists at the polls in Ireland in the British general election of 1918. Before that time other, more limited solutions had been the majority opinion among Irish people and also among Irish Catholics, if one would prefer this category as a more realistic image of the Irish nation.

One sometimes gets the impression that, now that the heyday of decolonization is over, there is again a more conscious search for more limited forms of self-government among those who look at themselves as nationalists if they only have limited resources and a moderate power base at their disposal. The limited success of many decolonized countries in terms of stable government and welfare growth may have resulted in a more relevant attitude towards full sovereignty by the latecomers. A case in point is the last remnant of Dutch colonialism, two triads of small islands in the Caribbean. Even those who consider themselves nationalists now resolutely refuse to cut the links with the colonial country that would be willing to do nearly whatever possible to bring this about.

But the Netherlands itself is in fact in a position that is not altogether different from this perspective. The recent political debate on the deployment of cruise missiles on Dutch soil brought in the open that the military structure of the NATO-alliance, in actual fact although not in principle, severely limits the autonomy of the Dutch government in security policy. This has been the case for some decades but it has largely been ignored. Even when the point was used in the political debate, it was apparently not the main point over which opinions clashed and it quickly returned to the background. In other words, if Dutch nationalism was still a significant element within the various doctrines that guide the Dutch parties, the aims of this nationalism do in fact no longer emphasize the fullest measure of political autonomy. This may again have to do with a recognition of the fact that full political autonomy is thought to be so disadvantageous that nationalists feel satisfied with more partial expressions of autonomy.

In more general terms it may be that full sovereignty as the singular acceptable aim of nationalism in the field of political autonomy has been the dominant version from roughly 1920 to 1970 but not necessarily before or after those years. This might mean that particularly outside the 1920-1970 period the assessments of the doctrine's fulfilment in terms of its own contemporary aims may well differ from an evaluation in terms of the one predetermined yardstick of full sovereignty.

If we want to evaluate nationalism's success in terms of the time it has taken to reach a certain point, we have to be sure about the time when nationalism started. This is difficult because of the uncertainty in determining threshold values for presence and because of the uncertain continuity of the nationalist movement. Nevertheless, it has been found for the period of decolonization that the period between the first signs of nationalism in a colony and the moment of independence became shorter as the process of decolonization generally progressed. Nijman (1985) has calculated that for twelve colonies where nationalism started before World War II it took 37 years on average to gain independence whereas for eleven countries with nationalism emerging after 1940, the mean number of years before independence was granted was 13.

There are two basically different yardsticks for the success of nationalism that do not necessarily go together. They are implied by the two aims that define the doctrine's contents: political autonomy and the realization of the nation. There is no reason why these two dimensions would necessarily go together. It is possible to think of cases where a moderate effort enabled a nationalist movement to acquire a high degree of political autonomy (it is even possible to acquire political autonomy without nationalist doctrine at all, but some rhetoric in that direction has apparently become inevitable). On the other hand even impressive shows of nationalistic mobilization of movements may not entail the realization of political autonomy.

The clearest examples of successes for nationalist doctrines with minimal effect have passed during the age of decolonization. Countries like Mauritania, Zaire, some Caribbean islands and some archipelagos in the Pacific have got full political autonomy with the tiniest nationalist movements one can think of, even taking their size into account. On the other hand we have clear shows of strength in terms of national mobilization not followed by a proportionate increase in political autonomy in Eastern Europe such as Hungary 1956, Czechoslovakia 1968, and Poland from 1980 onwards, not to speak of earlier examples in that country.

There are of course also cases where the acquisition of political autonomy and the realization of the nation tend to go together. The fiercest struggles of the French against the nationalist resistance in their colonial empire induced an extraordinary rate of mobilization in terms of numbers of people involved and resources per person committed. Indo-China, particularly Vietnam, and Algeria suffered terribly from the clash. Finally, in Vietnam after yet another war, full political autonomy for the whole territory was granted. Nationalism as the prime moving force of the leading unit became more and more immersed in communism. At the beginning of World War I a number of states that already had political autonomy were able to mobilize and thus to realize their respective nations to an unprecedented degree. The crystallization of a diversity of viewpoints in well-differentiated party systems had for a time been underway, but all heterogeneity was homogenized at one stroke when war credits were voted nearly unanimously in the national parliaments. This was the evocative symbol of the end of internationalism of some of these parties.

If political autonomy and the realization of the nation are the yardsticks for success once nationalism has got started, what are the conditions for success in these two dimensions? A large part of the provisional answers I have is implied by what has been said. If the assets of the leading unit, a historical core, mobilizational capacities

in society at large, a lack of powerful alternative ideologies and triggering factors like threats are conditions to
bring nationalism to life, the same conditions, at a higher
level only, determine its success in terms of the realization of the nation.

On the other hand, there is no such relationship with
the acquisition of political autonomy, large-scale or
small-scale. The acquisition of political autonomy is a
function of the power nationalism is able to generate. This
is by no means a question of numbers only. A small,
highly-motivated, ruthless, technically-competent group may
be able to wring political autonomy from the hands of a
contender and impose it on the population at large, pretend-
ing it is in fact operating in the nation's name. The
Russian revolution and the takeovers it inspired may do by
way of examples. They are curious mixes of nationalism and
communism.

Present-day terrorism seems primarily intent to keep
its nationalist aims before the public and to arouse
interest by acts of spectacular, highly symbolic violence
(Schmid, 1984). This is not to say that the violence used
in terrorism is less cruel and gruesome than violence used
in other settings. It is simply to argue that the major
intentions of terrorists, at least of the framers of these
acts, are in the messages the violence conveys to others. As
far as a direct assault on the state in order to gain
political autonomy is not really at issue, this form of
terrorism is more aiming at nation-building than at state-
making. The doctrines of terrorists are not nationalist by
definition, but many have at least nationalist overtones.

The acquisition of political autonomy is also, more
stringently than the realization of the nation, a function
of the willingness of other parties to accept it. Therefore
the values, utilities and ensuing preferences of well-organ-
ized other parties (e.g. states) should be taken into
account in this instance. So whatever the degree of
mobilization practically all colonies got full sovereignty.
It may be added that the sequence in which they got it was
to a considerable degree determined not only by available
intelligentsia, mobilizational capacity etc., but was also
clearly affected by the number of white settlers, the nature
of government, the economic strength and the links with the
US or the mother country (democracy, economic strength and
friendly relations with the US seemed to spur decolonization: Nijman, 1985). Whatever the degree of mobilization
in Eastern Europe it is only weakly reflected in more
political autonomy. In Ireland considerable autonomy to a
major part of the island (the Irish Free State) was granted
on account of a far more ruthless campaign, probably less
broadly supported than the preceding ones, for a full
measure of autonomy, but also because the British were more

willing to consider such proposals given the state they were in after 1918 and the international climate of opinion on self-determination etc.

Finally, we may consider the case of a fully-satisfied nationalism fallen asleep to such a degree that it may never awaken again. In a few countries like the Netherlands this might be the case. We will have to look for signs that the importance of political autonomy for something called the Dutch nation is still relevant and that this potential nation might again be mobilized if the need for it was felt by some leading unit. The facts about the abrogation of an independent security policy already mentioned plus the ever-more important economic links with partners elsewhere, particularly in the EC, undercut the idea and the credibility of political autonomy to a significant degree. This, however, may give rise to a more moderate nationalism only. But one can possibly argue an even stronger position.

The relation to the country as a whole is, I presume, increasingly perceived as a business-like exchange relation to the state that should be conducted in a tit-for-tat manner. Not much primary value-laden attachment to a Dutch nation is being shown. Flags seem more rarely to be put out than they used to. The national anthem is not often heard. We will, however, have to wait for the next royal wedding to be surer. If nationalism in this case would eventually falter, wane and disappear it is by no means clear what will come next. No trace of a nationalism at the European scale is to be found, nor can I see the Friesians and other even more insignificant regional movements gaining in strength. This eventual failure of nationalism may induce a search for new doctrines, the outcomes of which are far too hazy to speculate about for the moment.

References

Deutsch, K.W. (1961) 'Social Mobilization and Political Development', American Political Science Review, 55, 493-515.

Hirschman, A.O. (1970) Exit, Voice and Loyalty, Harvard University Press, Cambridge Mass.

Hobsbawm, E. & T. Ranger (eds.) (1984) The Invention of Tradition, Cambridge University Press, Cambridge.

Nijman, J. (1985) De Onafhankelijkheidsdatum. Een explorerend onderzoek naar de determinanten van het tijdstip van politieke dekolonisatie. Unpublished thesis, University of Amsterdam.

Schmid, A.P. (1984) Political Terrorism. A research guide to concepts, theories, databases and literature, North Holland Publishing Company, Amsterdam.

Smith, A.D. (1981) 'War and Ethnicity: the Role of Warfare in the Formation, Self-Images and Cohesion of Ethnic Communities', Ethnic and Racial Studies, 4, 375-397.

Tilly, Ch. (ed.) (1975) The Formation of National States in Europe, Princeton University Press, Princeton.

Williams, C.H. (1986) 'The Question of National Congruence', In. R.J. Johnston & P.J. Taylor (eds.) A World in Crisis? Geographical Perspectives, Basil Blackwell, Oxford, 196-230.

Wusten, H. van der (1977) Iers verzet tegen de staatkundige eenheid der Britse eilanden 1800-1921. Een politiek-geografische studie van integratie - en desintegratie-processen. Amsterdam (Ph.D. Thesis summarized in Netherlands Journal of Sociology 1980, 2).

Chapter 13

MINORITY NATIONALIST HISTORIOGRAPHY

Colin H. Williams

'Tell it as it is' is a favourite adage of oppressed peoples
faced with the task of creating their version of historical
events. A less favourite observation for them is that there
are at least two sides to any story. The version presented
by the powerful usually becomes accepted as received truth,
that presented by the powerless in time fades into oblivion;
or more accurately into the reference and footnote section
of academic and school texts as interesting but misinformed
and partial accounts of past events. In this chapter I will
argue that in the main geographic forays into the field of
nationalism have tended to tell the victor's side of the
story and have left the vanquished to cry in the wilderness.
This is an understandable, if regrettable, tendency for it
is often safer and more academically respectable to cite the
views and interpretations of leading political and social
actors from the dominant state culture than it is to enquire
into the world views and priorities of representatives of
the subordinate culture(s). My specific concern is to
highlight two tendencies which serve to gloss over the
origins of selected nationalist movements and to reduce the
significance of nationalist historiography in our reconstruction of the past.
      The first tendency, which is common in geography, is
to generalise so much about the core doctrine, values and ·
political orientation of nationalist movements as to lose
not only the historical accuracy of the analysis but also
the dynamism and conflict inherent in the developing
movement itself. As a counter to this tendency I will argue
that we should emphasise the significant role of key
nationalist actors.
      The second tendency is the acceptance of the dominance
of a state-centred, often core-periphery, perspective and
its related set of assumptions. It can induce a pejorative
statist framework of analysis which, at times, allows little
free reign for the activities of the minorities under
scrutiny. They are often labelled and treated as vic-

tim-like, reactive agencies, rather than as purposive dynamic and creative individuals and groups. Indeed for a discipline with a long tradition of analysing self-determination, emphasis on the 'self-hood' of groups, let alone their 'determination', is often threadbare. This perspective also gives primacy to the 'output' side of the relationship between state and nation(s). It has several implications which merit serious consideration:

a)   it marginalises the aims and values of the subordinate group;

b)   it leads to an over-concentration on the role and performance of political parties in state-wide elections, and to an often consequent misinterpretation of the significance of nationalist movements, whose fortunes are measured electorally once every few years;

c)   it encourages us to be overly concerned with assessing the effectiveness of the state and its machinery in its response to the nationalist challenge; and

d)   it leads to a comparative analysis of the state, rather than to a comparative analysis also of nationalist movements, and when such movements are compared they tend to be discussed in abstract, universal terms, rather than in detailed, place-specific terms.

The last point should not be misinterpreted as an attack on universal theory, but rather as a realisation that such theory should be based on a wide range of detailed knowledge, as was illustrated in Orridge and Williams (1982).

There are understandable reasons for this trend. Geography's epistemological background and recent methodological training conduce to an aggregate, often statistical form of analysis. Whilst this emphasis certainly adds a distinctive perspective to the now-voluminous literature on nationalism, it also underplays the unique and the idiosyncratic forms of nationalism. In addition, geographers, if they are bi- or tri-lingual, are more likely to read and report on the literature and political decisions of dominant, often state-controlling, groups rather than on their corresponding aspiring nationalist movements, e.g. Spanish rather than Euskera or Catalan, French rather than Breton. This leads to an under-representation of minority nationalist ideas and actions in the international scholarly literature. When such affairs are translated into an international language their poignancy is often muted by their rendition into the 'majority tongue'. It is not incidental that the general power-relationship between dominant and a 'subordinate' language, to which this paragraph alludes, is also an aspect of the legitimisation of ideas and of relative group positions in the wider society (Williams, 1984a). In consequence we can learn much about nationalism if we pause, and take stock of the significance of nationalist historiography which attempts to reverse these tendencies.

I have argued elsewhere that a sine qua non of nationalist movements is that their leaders share a sense of distinct history, a sense of belonging to a unique place, and a sense of their own mission/destiny, which animates their political drive (Williams, 1984b). These three axes define the elementary historiography of most movements. In this chapter I want to examine these three senses, illustrated by a leading Welsh nationalist theoretician. This concern with historiography goes against general trends in two respects: a) it is concerned with idealist rather than with materialist issues and analytical forms; and b) it concentrates on the 'input' side of nationalism, and ignores (for the present) the 'output' side, that is, the state's response and the subsequent impact of the movement on either its target population or on the wider political stage (Butt-Philip, 1975; Morgan, 1981; Williams, 1982; Davies, 1983).

History is vital to the nationalist, as to all who would seek to change the existing socio-political order. But history, as many scholars would remind us, needs to be re-told, re-cast and re-created in order to serve the present. Williams and Smith (1983) illustrated how nationalists have appropriated both the past and the contemporary landscape in order to reconstruct a social order in tune with a nationalist vision of the world. More recently, Smith (1986), in expanding on this re-creation of nations as a continuous task, has written:

> Creating nations is a recurrent activity, which has to be renewed periodically. It is one that involves ceaseless re-interpretations, rediscoveries and reconstructions; each generation must re-fashion national institutions and stratification systems in the light of the myths, memories, values and symbols of the 'past', which can best minister to the needs and aspirations of its dominant social groups and institutions. Hence, that activity of rediscovery and re-interpretation is never complete and never simple; it is the product of dialogues between the major social groups and institutions within the boundaries of the 'nation', and it answers to their perceived ideals and interests. (p. 206)

In the mid-nineteenth century the signal appeal of non-state nationalism was that it promised to introduce a hitherto under-represented section of the community into mass politics, namely the cultural intelligentsia. They had recognised that the identity of a nation was bound up with individual and collective memory, and sought to engage history as an ally in the political struggle for independent statehood. In historically autonomous states such as France and Britain, the historical continuity of the nation-state was expressed primarily through legal and political institutions which embodied their national culture. But for peoples without a separate institutional state apparatus, such as the Germans, Welsh, Slavs, Basques and Bretons, this attempt to make state and nation conterminous was not possible (Williams, 1984b; 1986). In consequence they had to exploit other resources and appeal to different norms and obligations, so that, as Berry (1981, p. 83) has argued, 'the grounding of temporal value' in political continuity being barred to them, they sought this ground and legitimacy in 'cultural continuity' and collective memory. A prime element in this form of identity is language and its associated territorial homeland. Nationalist historiography was to exploit this base and to construct an alternative interpretation of the role of their nation in state-building from that offered by the centralist state historians.

Religion was a central plank of some nationalist theorising. Tracing one's cultural lineage to the Judaeo-Christian origins of European civilisation was an obvious means of legitimising one's national past in a common heritage. In this chapter I will focus on Christian idealism as a source of socio-political action for it figures prominently in the arguments and theories of prominent nationalists such as Sabino de Arana, Saunders Lewis and Daniel O'Connell (Hill, 1980). Particular attention is paid to the Welsh experience, as it is the case that I know best; I would argue it is reflective of other European minority situations, particularly those on the Atlantic seaboard (Williams, 1986).

A Divine Heritage

For well over three centuries, and especially since the age of mass literacy and cheap pamphlets and newspapers, the Welsh intelligentsia have likened their people's predicament to that of the Old Testament Jews. The sons and daughters of Abraham and Sarah were living representatives of the original nation whose trials, tribulations and temptations, even today, are fertile ground for the allegorising prophets of Nonconformist Gwalia.

Such was the deep impact of Biblical scholarship that the 'werin' (common people) tended to know more about Hebron and Hermon that they did about Halifax or Huddersfield. They were more alive to the ancient needs and misdemeanours of the Amorites, Chilmadians or Hittites than they were to the very existence of the Croats or Moravians. Their God, however, was a partial being when it came to dispensation and grace. Undoubtedly Yahweh's identification with the suffering and the oppressed, His 'kinship for the poor' as expressed in the Psalms (72:12; 113:7; 146:7 and 9) was balm indeed for those overburdened with low prices and high rents, with poor soil and an uncertain tenancy subject to the whim of an alien, landowning class of gentry, whose twin institutional pillars, the Crown and the Church of England, were likened unfavourably with the power of Imperial Rome and the legalistic Sanhedrin of old. However, God was also just, and when His people disobeyed, then judgement was swift. The times of His elect and of the nations were numbered, they would know that His 'justice would roll down like the waters, and righteousness like an everflowing stream' (Amos, 5:24). The Welsh were in judgement, their land despoiled, their culture and language threatened with extinction, and their yielding to the temptations of Anglicised modernisation was an unforgivable sin. The old cliche rang out from every Tabernacle and Siloam, Wales was 'so near to England, so far from God'.

Geographically, the Welsh were portrayed as a modern version of Palestinian Jews. They too were rural pastoralists who inhabited a rugged, mountainous terrain, watered by swift streams and tempestuous seas. The territorial dimension of Judaism, brilliantly analysed by W.D. Davies, in a series of significant works (1974; 1982), was shadowed closely in the indissoluble bond between God, the people and the Welsh landscape, which Professor J.R. Jones termed 'cydymdreiddiad'. It was a resonant sentiment which proved a powerful fillip to the Welsh Liberal and Cymru Fydd cause of Home Rule. Claims of fulfilling 'the desire of nations' fell on receptive ears in the latter part of the nineteenth century and the pre-war lull. The nonconformist press, its politicised leadership and the myriad radical discussion groups were vital channels for the formulation and dissemination of nationalist ideas which placed Wales, its culture and economic problems at the forefront of the political agenda.

Out of this crucible was formed a generation of political leaders and intellectuals. However, it would be incorrect to assume, as many do, that the leaders of Plaid Genedlaethol Cymru were a homogeneous group, with essentially similar social backgrounds and political ideals. Within the aegis of Christian values there emerged very sharp differences of interpretation as to what should be done to 'save Wales'. This chapter presents elements of these major interpretations, conscious that the early movement was not primarily concerned with political independence, but rather with the preservation of a particular socio-economic conception of a rural, communitarian, and essentially Welsh-speaking society.

A European Moral Perspective

A major concern of early Welsh nationalist thinking was the re-definition of the political order in moral not material-ist terms. Historicism was a relatively youthful Hegelian trend which Welsh thinkers were beginning to employ to legitimise their social and cultural formulations, but whereas many scholars turned to the Celtic realm for moral inspiration and putative 'national' traits, Saunders Lewis sought his authenticity in the Catholic, Latin civilisation of Europe. Medieval Europe possessed a unity of spirit and of law which nurtured minority cultures because diversity could be accommodated within a universal European civilisa-tion. In his seminal paper 'Egwyddorion Cenedlaetholdeb' (Principles of Nationalism) delivered at Plaid Cymru's first Summer School in 1926, Lewis outlined his conception of Welsh national history, an interpretation which in broad terms is still current today amongst nationalists. I have presented a schema of his ideas in Table 1, and in this discussion will refer only to the salient points, which contribute to a nationalist world-view.

Lewis, the principal early theoretician of the national movement, compares and contrasts two types of nationalism: a) state nationalism, which emerged in the sixteenth century to challenge the universal moral order of the Church; and b) organic nationalism, a revised version of an earlier form of political doctrine developed under medieval Christendom. He argued that his compatriots should rediscover this pre-existent nationalism, to counter the legitimacy of British state nationalism. He asks:

> What then is our nationalism? It is this: to return to the principle accepted in the Middle Ages: to repudiate the idea of political uniformity, and to expose its ill-effects: to plead therefore for the principle of unity and diversity. To fight not for Welsh independence, but for the civilisation of Wales. To claim for Wales not independence, but freedom. And to claim for her a place in the League of Nations and in the community of Europe, by virtue of her civilis-ation and values (Lewis, 1926; Jones and Thomas, ed., 1973, p. 29)

The keystone of Welsh civilisation was the defence and promotion of the Welsh language, a theme which came to dominate many of the activities of Plaid Cymru, 1925-1974 (Williams, 1982). Lewis justified its promotion in moral as well as in social terms; it was a testimony of Wales's continuing cultural contribution to a common European order. The language was proof of the Welsh having maintained the faith, so to speak, of traditional European values. Lewis's continentalism was born out of his admiration for classical literature, for several French Catholic contempories and, of course, from his own conversion to Catholicism (Lewis, 1986; Jones, 1986).

In a trenchant analysis of Lewis's political thought, Dafidd Glyn Jones (1973) argues that although Lewis's interpretation of Welsh national history and destiny did not square with mainstream Liberal nonconformity, it was nevertheless an influential thesis. In brief, Lewis counterposed the twin external forces acting upon contemporary Wales, the Empire and the League of Nations. He urged fellow nationalists to shun the Imperial dream and work to re-construct a Europe of the Nations in which Wales, along with the other British nations, would be represented as a free, democratic country. In truth the prospect of a quasi-federal Europe at this time was idealistic in the extreme, but the lessons for the future political ideology of small nations are resonant still. Lewis links the Welsh with a continental European tradition, by-passing the English, Westminster and paganism. His discourse reveals his concern to distance Wales from British imperialism and the destructive consequences of state integration which had daily sealed the populace within an apparently inexorable process of national congruence since the era of sixteenth century nationalism (Table 1) (Williams, 1986).

Table 1. A Summary of Saunders Lewis's Interpretation of Welsh History, derived from his Egwyddorion Cenedlaetholdeb; (Principles of Nationalism), 1926.

ROMAN EUROPE
Condition   Unity enforced by a dominant Christian, Latin civilization which induced a European moral integrity.
Result   Minority peoples, though conquered, were elevated by sharing a powerful civilizing tradition.

MEDIEVAL EUROPE
Condition   The one and indivisible Church exercising supranational authority.
Result   Local cultural diversity nurtured and protected within a framework of spiritual and legal unity.

REFORMATION EUROPE
Condition   Successive challenges to the universal Christian
             order by individual and institution-
             al interests, e.g. Luther, Machia-
             velli and the Tudor monarchs.
Result   Church authority denied; the King's sovereignty
             and writ established; the state now
             replaces the Church as the supreme
             sovereign body, leading to confu-
             sion, disintegration and the genesis
             of state's drive towards unitary
             principles.

16 C. NATIONALISM
Condition   Unification and integration sought within states
             by establishing one government, one
             language, one state law, one
             culture, one education, one relig-
             ion.
Result   'The triumph of materialism over spirituality of
             paganism over Christianity, of
             England over Wales'.

CONTEMPORARY BRITAIN (1926)
Condition   Imperialistic and Marxist challenges within an
             advanced industrial order.
Result   A materialistic spirit of narrow and godless
             nationalism; destroying the indi-
             viduality of Wales.

PRESCRIPTION   The establishment of a central Welsh
                authority, exercising self-govern-
                ment and guaranteeing the primacy of
                the Welsh language in all aspects of
                public life. This would engender 'a
                generous spirit of love for civili-
                zation and tradition'.

In contradistinction to a traditional Whig version of
the historical development of the British Isles which
portrayed the Celtic remnants as vanquished barbarians
driven ever westward by powerful Germanic tribes, Lewis
sought to harness the moral and spiritual legitimacy of the
Imperial Roman inheritance and embody its finer virtues
within a surviving British culture. Clearly this was a
twentieth century ex post facto justification for the
virility displayed by Celtic culture, in its tenacious hold
on the western margins of these isles, but it was also a
simultaneous rebuttal of the inherent superiority of

Anglo-Saxon culture and politics as represented in the contemporary 'English' state and its repressive actions in continuing to deny the appeal of Celtic freedom.

> He chose to regard the Welsh, in their origin, not as a people driven headlong to the West and the mountains before a swift and irreversible Anglo-Saxon onslaught, but as that section of the Britanni who, during and after the Roman withdrawal, opted to identify themselves with the cultural and spiritual ideal of Romanitas, which by then included the Christian religion (Jones, 1973, p. 33).

Here is the original thesis:

> But is there a European tradition to be found within Britain? Is there here a nation which was, in its origins, part of Western Civilization, which thinks in the western way, and which is able to understand Europe and sympathise with her? The answer is: Wales. The Welsh are the only nation in Britain who have been part of the Roman Empire, who, in childhood, were weaned on the milk of the West, and who have the blood of the West in their veins. Wales can understand Europe, for she is one of the family. If a choice must be made, as Chamberlain insists, between the Empire and the League of Nations, there can be no doubt as to which way Wales will tend. To her, always, and to the greatest of her sons in thought and learning, contact with Europe has meant a renaissance and an inspiration. To her, the Empire was never anything but a name and an empty noise ... This, then, is the reason why she must demand a seat in the League of Nations, so that she may act as Europe's interpreter in Britain, and as a link to bind England and the Empire to Christendom and to the League itself (quoted in Jones, 1973, p. 33).

Despite inducing internal division and strain within the nationalist movement, especially with regard to allegations of his sympathy with quasi-Fascist organisations, elements of Lewis's interpretation of the correct relationship between Wales and Europe and of the Welsh role as 'Europe's interpreters in Britain' informed subsequent nationalist thought. Jones (1973) cites as evidence Wade-Evans's opening lecture in the symposium the Historical Bases of Welsh Nationalism (1950), and Gwynfor Evans's book Aros Mae (It Still Remains), together with other of Lewis's later works, e.g. Buchedd Garmon (The Life of St. Germanus) which dramatises the conflict between Romanitas and Barbaritas in fifth-century Britain.

In this re-orientation of Welsh politics and history Lewis sought to present a mutually exclusive set of alternative identities to his audience. One could either be 'truly Welsh' and uphold the national vision of Wales being a co-equal partner in European civilisation and development, or one could deny this vision and accept the English position that the United Kingdom was an indivisible state wherein English values and priorities took precedence. It was to be an uphill struggle, for despite two centuries of seeking to create an autonomous Welsh identity, the romantic mythologists had yet to permeate the popular consciousness in political terms (Morgan, 1983).

Lewis's social policy was predicated upon his belief that communities were the sine qua non of civilisation, they were the organic, historically transcendent reference group. In Canlyn Arthur (1938) he suggests that the nation is composed of a community of communities which acts as a filter between the individual and the state,

> Family and tribe existed prior to the state, and voluntary organisations existed prior to the authority of sovereign government ... A nation's civilization is rich and complex simply because it is a community of communities,and for that reason also the freedom of the individual is a feasible proposition ... His liberty depends upon his being a member not of one association but of many (quoted in Jones, 1973, p. 34).

'Social pluralism' was to be encouraged so as to stifle the influence of humanitarianism and state-centralism which eroded the family, its social responsibility and a Christian moral order.

During the depression years Lewis's conception of nationalism took on a more radical hue, especially with respect to the task of social and economic reconstruction. In May 1932 he declared that:

> The hideousness grows nakeder in these times of distress. And if I am asked what is the essential raison d'etre of Welsh Nationalism, I answer that first and indisputably first it is to change the entire system of government and of imperialist capitalism that has made my country the worst hell in Europe today. (The Welsh Nationalist, May 1932, quoted in Davies, 1983, p. 100)

He advocated a revolutionary programme of social change which would displace the barbarism and horror of industrial capitalism with the 'Welsh common tradition', a shadowy but nevertheless powerful concept which was enunciated in his

Ten Points of Policy of 1934 (Table 2). These points were
his summary of the social teaching, or 'social catechism',
of Welsh nationalism (Davies, 1983, pp. 100-102), a reflection of his increasingly self-conscious attempt to give a
practical edge to Christian sociology.

Table 2: The Ten Points of Policy

1. It is not the function of government to create a
   complete system and economic machine that people
   accept and conform to. The function of government is
   to create and sustain energetically the conditions and
   circumstances that will provide the opportunity,
   leadership and support for the nation itself to
   develop the system that will be most congenial to its
   ideals and traditions and that will be the means to
   ensure the health of society and contentment of
   individuals.

2. The economic unit, as far as it is possible, should be
   at one with the political and social unit, for only in
   that way is it possible to defend people from external
   oppression. Neither should Wales be so dependent for
   its sustenance on other countries that it would be
   unable, if war or some other necessity occurred, to
   feed its own people and keep them from hunger.

3. That industrial capitalism and economic competition
   free from the control of government (i.e. free trade)
   are a great evil and are completely contrary to the
   philosophy of cooperative nationalism.

4. That it is part of the task of a Welsh government to
   control money and conditions and credit institutions
   for the benefit of industry and social developments.

5. Trade unions, works committees, industrial boards,
   economic councils and a national economic council,
   cooperative societies of individuals and of local and
   administrative authorities, should have a prominent
   and controlling role in the economic organisation of
   Welsh society.

6. The families of a nation should be free, secure, and
   as independent as possible. To enable that it is
   necessary to legislate and plan substantial distribution of ownership, because only a man with property
   can be a free man. Ownership should be distributed so

widely among the families of the nation that neither state nor individual nor a collection of individuals can oppress the people economically.

7. Agriculture should be the chief industry of Wales and the basis of its civilisation.

8. For the sake of the moral health of Wales and for the moral and physical welfare of its population, South Wales must be de-industrialised. All the natural resources of Wales are riches to be dealt with carefully for the benefit of the Welsh nation and for the benefit of its neighbours in other parts of the world.

9. No right or unqualified ownership will be recognised unless linked with social duties and responsibilities.

10. It will be part of the function of a Welsh government to cooperate with other governments with regard to the problems of provisions and industrial organisation. Freedom and encouragement will be given to trade unions and Welsh industrial boards to cooperate and consult with equal and similar unions and boards in other countries or by means of the International Labour Office.

Source: Y Ddraig Goch, March,, 1934; also 'Deg Pwynt Polisi' Canlyn Arthur, Gwasg Aberystwyth, 1938, pp. 11-13; reproduced in D.H. Davies, The Welsh Nationalist Party, 1925-1945, Cardiff, 1983, pp. 100-01.

A detailed critique of the influence of the 'Ten points' on the development of Plaid Cymru's philosophy may be found in Jones (1973) and Davies (1983). The social philosophy they espouse has been criticised for being no more 'than the social teaching of the Roman Catholic Church' into which Lewis had been received on 16 February 1932 (Davies, 1983, p. 102). Certainly his writings equated Welsh nationalism with Christianity, primarily because he valued the individual and spiritual element of civilisation in contrast to the corporate and materialist claims of both Fascism and Marxism. The latter competitive ideologies were both dismissed for alienating individuals and their constituent communities from their 'real' past. Class analysis was misguided because by obscuring the relationships between the individual and his/her Maker, and between brother and sister, it stifled the outworking of Christian values in society. Idealist, moral convictions were not only central in thought, they were purposively promoted in action, for some of his nonconformist colleagues recognised the 'Ten Points' as the modern expression of 'traditional Welsh social philosophy' (Davies, 1983, p. 100). Lewis's own conviction was uncompromising.

> I believe it is accurate to say that Christian motives greatly influenced the formation of the party, that it has sought to base itself on Christian sociology, and that Christianity is as essential to the Nationalist Party as is anti-Christian materialism to Marxism. (S. Lewis, Y Ddraig Goch, March 1938, quoted in Davies, 1983, p. 102)

Unfortunately for Lewis, his variety of Christian doctrine, Catholic and continental, did not engender much sympathy amongst many of his erstwhile supporters and party delegates, some of whom called for his resignation as chairman of the party (Jones, 1970; Jenkins, 1981). Although Lewis does not mention this issue in his letter of resignation to J.E. Jones, the Party Secretary, in May 1939, there can be no doubt that he felt his Catholicism might have contributed to the lack of cooperation and feeling of mistrust engendered in some quarters. Lewis's critics have been dismissed by the former Party Secretary as being opportunist and unrepresentative of the nonconformist majority (Jones, 1970, p. 205). Others, who equated support for European Catholicism with sympathy toward Fascism in Italy and Spain, overlooked the fact that many of Fascism's opponents in Germany, Italy and Euskadi were also Catholic (Jenkins, 1981, p. 63).

The fact remains, however, that the president and a handful of influential activitists at the core of the movement did espouse Catholic-derived policies. D.G. Jones (1973) speaks of two Papal encyclicals being adopted wholeheartedly (i.e. Rerum Novarum, 1891, and Quadragesimo Anno, 1931). It was not until the post-war period that the old dissenting denominations, the Baptists and Independents, came to dominate the party and mould its policies more in keeping with the radical-Liberal traditions of the Welsh-speaking gwerin/proletariat.

A Sense of Place

If Saunders Lewis gave to Welsh nationalism a sense of unbroken historical affiliation with mainstream European civilisation, J.R. Jones gave it a sense of place by rooting the national community in the Welsh landscape.

Like Lewis his starting point is the community and his analysis of Welsh national identity continues to animate nationalist thought. In his book Prydeindod (Britishness, 1966), he draws a significant distinction between a people and a nation: a people he defines as the product of the interpenetration of a land and a language, a process he calls 'cydymdreiddiad tir ac iaith'; a nation he defines as a tripartite relationship of land, language and state sovereignty. 'Cydymdreiddiad tir ac iaith' is central to the creation of a people, it has two formative bonds: a) an external spatial impulse - the territorial realm within which they have co-habited over the centuries; and b) an internal spiritual impulse - the language which transmits a people's heritage and becomes the repository of its culture. His thesis seeks to answer three questions: a) how does language bond? b) how does territory bond?; and c) what does 'cydymdreiddiad tir ac iaith' actually constitute?

Language bonds a people by acting on their spirits and through their collective memory by their blood ties. He is anxious not to appear to be arguing for a Fascist position based upon the purity of blood descent, but rather that language shapes a collective consciousness (shades of Heidegger here), although he attributes his understanding of this to Fichte without necessarily adopting the organic implications of Fichtean thought.

Territory bonds because of a universal need to be rooted, to be anchored in space. Jones claims that a thousand secret chords bind people to the land, and centuries of continuous occupation confirm the land as the vessel safeguarding and nurturing all cultural traditions. The interpenetration of land and language is not a natural process. It is an experiential act which reproduces itself daily in people's souls and therefore collectively is witnessed in society's commonplace acts: in the naming of mountains and vales, rivers and villages. Language is the key to meaning and identity; consider Jones's imagery:

> In this marriage and, as its foundations ... we see People, as it were, taking hold of their land and partnering it into the texture of their lives through the intercession of language. They would, as it were, see and handle and love the earth through the mirror of their language (Jones, 1966, p. 14, my translation).

When he turns his attention to the reality of Wales, Jones laments the scene of region after region losing its vital Welshness. All that remained in many areas was a relict landscape where only the toponymic features evidenced the once dynamic process of interpenetration of land and language, of settlement and dependence. Were people to forget past landscape creation they would lose their ability to recreate their present identity. For identity was not a given, it had to be made, remade and defended in the land. If successive areas of Wales, such as Tryweryn and Gwendraeth, were yielded or re-designated as part of a wider British space, to be appropriated as overspill towns or as water supply areas, then the link between territory and a specific people would be denied and they would be re-defined in functional terms as members of another nation. He cites Tillich's evocative reasoning:

> Being means having space or, more exactly, providing space for oneself. This is the reason for the tremendous importance of geographical space and the fight for its possession by power groups. The struggle is not simply an attempt to remove another group from a given space. The real purpose is to draw this space into a larger power field, to deprive it of a centre of its own. (P. Tillich, <u>Love, Power and Justice</u>, p. 100).

However, Jones exclaims that there never was a powerful centre to Welsh space, in consequence Tryweryn was seized, and when Midlands overspill new towns are accepted, the Welsh demonstrate their unworthiness to be called a nation.

A critical problem of his formulation of interpenetration leading to peoplehood, is the question which animates many of the 89 per cent in Wales who do not speak Welsh. If interpenetration is critical for separateness, what can be said of the condition of the anglicised Welsh person? Does he/she cease being Welsh? This was, of course, one of the great stumbling blocks to the development of the nationalist movement, which because of its early concern with rural, cultural affairs alienated the English-speaking, industrial populace of the north-east and south. Welsh identity, he reasons, is a continuum, many-layered and uneven in its distribution and effect. But by virtue of their residence all Welsh people have a consciousness of themselves as a separate people; some exercise this through the language, others hold the potential of realising this full identity when layer upon layer of Welshness can be exposed and adopted should the individual so will.

Though the Welsh are a people by his definition, they are not a nation, for they do not fulfil his requirement of there being a tripartite relationship between land, language and sovereignty. Only the English in these isles are a nation. The acid test of this assertion is the question of Britishness. Ask any English people what being British means and they will answer that it is synonymous with being English; there is no conflict of interest or of identity, for it is their state which has become the British state. No other nationalism competes for their loyalty. Not so the Welsh! For so long as the state has to deal with two identities in Wales, so long will the Welsh as a people survive. Once the English language and culture triumphs overall so as to leave only an echo of a former Welsh pesonality, then the English nation will have secured the Welsh territory in the British state (Jones, 1966).

British ideology persists, Jones argues, because it is a necessary means of dealing with the relative lack of state integration in Wales. It is the Welsh language which symbolises the separateness of the people; without it they would have been absorbed centuries ago into the English nation. Britishness is a myth, a self-deception that is pernicious because it is reproduced unconsciously.

> The creation of ideologies says Tillich, occurs unconsciously. It is not a conscious falsification, or a political lie. If this were the case, ideologies would not be very dangerous. But they are dangerous precisely because they are unconscious and are, therefore objects of belief and fanaticism (Jones, 1966, p. 39).

As Britishness is false-consciousness it follows that for Welsh people it can never provide the basis of a national identity. It does not reflect the intimate interpenetration of land and language and is an ideology not of their own making.

The consequence of this ideology is to induce among the population a split personality and a false premise for being, for existence. We should not forget, he argues, the experience of other Celtic peoples who forsook this act of interpenetration in Cornwall. For they were 'Britishised out of existence' - 'Fe'u Prydeiniwyd allan o fod'. (p. 43). A further consequence is to be consigned to a parasitic role on someone else's identity-formation process. The task facing the nationalist is to dissestablish British con-sciousness in Wales, for the end of Britishness could be the re-birth of Welshness. His work ends with a series of appeals to the politicised youth to protect their separate-ness, to insist upon their Welshness over and above the attractions of a British identity and to fight for their claim for nationhood. 'Eich yspryd yn unig a'ch cyfyd ac a'ch gesyd i sefyll ar eich traed' (p. 59).

Together with Saunders Lewis's famous radio broadcast, 'The Fate of the Language', Jones's call to action spurred on the Welsh Language Society of which he was the leading intellectual voice for the remainder of the decade.

Conclusion

The contribution of the intelligentsia to the formation and subsequent development of Welsh nationalism is well documented (Morgan, 1981; Davies, 1981b, 1984). I have chosen to emphasise the key conceptual contributions of two of the leading theorists for I am convinced that their influence has not been sufficiently realised outside Welsh-medium literature. But I am also convinced that the problems confronting both Lewis and Jones in impressing upon an unresponsive population the merit of a nationalist vision of socio-political change are common to many other European cases, where class politics and the drift towards decentralist socialism are more convincing bases for political mobilisation. For a major problem of Welsh nationalism has been its consistent under-representation of the significance of economic and social class issues in mass politics, a feature which has been redressed in the past decade. Plaid Cymru's emphasis on rural agrarian values did little to attract the attention, let alone the support, of the industrial proletariat. Its leadership represented the interests of small-town nonconformism, whose conception of a re-fashioned 'Europe of the nations' was far too idealistic in the 1930s, but may nevertheless achieve significance in the twenty-first century. Concern with community-level politics and with a preservation of a communatarian economy is currently re-echoed in the alliance between nationalists and other social movements such as the ecology and anti-nuclear movements, where anti-state rhetoric is combined with thoughtful analysis of local problems. We recognise that timing is all in politics. Nationalist historiography may have failed to produce an independent Wales, but it has surely contributed to the recognition that Wales is a separate place, wherein one may no longer say with impunity, as did the Encyclopedia Brittanica of old 'For Wales, see England'.

References

Anderson, B. (1983). Imagined Communities. Verso, London.
Blaut, J.M. (1986). 'A Theory of Nationalism', Antipode 18, pp. 5-10.
Breuilly, J. (1982). Nationalism and the State. Manchester University Press, Manchester.
Butt-Philip, A. (1975). The Welsh Question. The University of Wales Press, Cardiff.
Cooke, P. (1985). 'Class Practices as Regional Markers: A Contribution to Labour Geography', in D. Gregory and J. Urry (eds.) Social Relations and Spatial Structures. Macmillan, London.

Davies, D.H. (1983). The Welsh Nationalist Party, 1925-45. University of Wales Press, Cardiff.

Davies, J. (1981a). The Green and the Red. Y Lolfa, Talybont.

Davies, J. (ed.) (1981b). Cymru'n Deffro. Y Lolfa, Talybont.

Davies, J. (1984). 'Plaid Cymru in Transition' in J. Osmond (ed.) The National Question Again. Gomer Press, Llandysul.

Davies, W.D. (1974). The Gospel and the Land. University of California Press, Berkley, 1974.

Davies, W.D. (1982). The Territorial Dimension of Judaism. University of California Press, Berkley.

Evans, G. (1982). Aros Mae. Ty John Penry, Swansea.

Giddens, A. (1985). The Nation-State and Violence. Polity Press, Cambridge.

Haddock, B.A. (1980). An Introduction to Historical Thought. Arnold, London.

Hill, J.R. (1980).'The Intelligentsia and Irish Nationalism in the 1840s' Studia Hibernica, 20, pp. 73-109.

Humphreys, E. (1983). The Taliesin Tradition. Black Raven Press, London.

Jenkins, D. (1981). 'Penyberth a'r Cyfnod Wedyn, 1936-1938' in J. Davies (ed.) Cymru'n Deffro. Y Lolfa, Talybont.

Jones, A.R. and Thomas, G. (eds.) (1973). Presenting Saunders Lewis. University of Wales Press, Cardiff.

Jones, D.G. (1973). 'His Politics' in A.R. Jones and G. Thomas (eds.) Presenting Saunders Lewis. University of Wales, Cardiff

Jones, J.E. (1970). Tros Gymru. Ty John Penry, Swansea.

Jones, J.R. (1966). Prydeindod. Llyfrau'r Dryw, Llandybie.

Jones, M.P. (1986). 'Yr Awel o Ffrainc', Y Traethodydd, Gorffenaf.

Lewis, S. (1926). Egwyddorion Cenedlaetholdeb. Plaid Cymru, Cardiff.

Lewis, S. (1937). Buchedd Garmon. Aberystwyth.

Lewis, S. (1938). Canlyn Arthur. Gwasg Aberystwyth, Aberystwyth.

Lewis, S. (1986). Ati, Wyr Ifainc. Gwasg Prifysgol Cymru, Cardiff.

Michel, N. (1986). Le Séparatisme en Bretagne. Les Bibliophiles de Bretagne, Beltan, Brasports.

Morgan, K.O. (1981). Rebirth of a Nation: Wales 1880-1980. The University of Wales Press, Cardiff.

Morgan, P. (1983). 'From a Death to a View' in E. Hobsbawm and T. Ranger (eds.) The Invention of Tradition. Cambridge University Press, Cambridge.

Orridge, A.W. and Williams, C.H. (1982). 'Autonomist Nationalism: A Theoretical Framework for Spatial Variations in its Genesis and Development', Political Geography Quarterly, 1, pp. 19-39.

Smith, A.D. (1986). The Ethnic Origins of Nations. Black-
     well, Oxford.
Wade-Evans, A.W. et al. (1950). Seiliau Hanesyddol Cened-
     laetholdeb Cymru. Plaid Cymru, Cardiff.
Williams, C.H. (1982). 'Separatism and the Mobilization of
     Welsh National Identity' in C.H. Williams (ed.)
     National Separatism. University of Wales Press,
     Cardiff, pp. 145-201.
Williams, C.H. (1984a). 'More Than Tongue Can Tell:
     Linguistic Factors in Ethnic Separatism' in J. Edwards
     (ed.) Linguistic Minorities: Policies and Pluralism.
     Academic Press, London, pp. 179-221.
Williams, C.H. (1984b). 'Ideology and the Interpretation of
     Minority Cultures', Political Geography Quarterly, 3,
     pp. 105-25.
Williams, C.H. (1986). 'The Question of National Congru-
     ence', in R.J. Johnston and P.J. Taylor (eds.) A World
     in Crisis? Blackwell, Oxford.
Williams, G. (1984). 'What is Wales?: the Discourse of
     Devolution', Ethnic and Racial Studies, 7, pp.
     138-59.
Williams, C.H. and Smith, A.D. (1983). 'The National
     Construction of Social Space', Progress in Human
     Geography, 7, pp. 502-18.

ABOUT THE AUTHORS

James Anderson, Faculty of Social Sciences, The Open University, Walton Hall, Milton Keynes, MK7 6AA, U.K.

Bertha K. Becker, Department of Geography Federal University of Rio de Janeiro, Rio de Janeiro, Brazil, South America.

Andrew Burghardt, Department of Geography, McMaster University, 1280 Main Street, Hamilton, Ontario L8S 4K1, Canada.

Josiah A.M. Cobbah, Urban Morgan Institute for Human Rights, College of Law, University of Cincinnati, Cincinnati, Ohio 45221, U.S.A.

Alasdair Drysdale, Department of Geography, University of New Hampshire, Durham, New Hampshire 03824, U.S.A.

Clive Hedges, Faculty of Social Sciences, The Open University, Walton Hall, Milton Keynes, MK7 6AA, U.K.

R.J. Johnston, Department of Geography, University of Sheffield, Sheffield S10 2TN, U.K.

David B. Knight, Department of Geography, Carleton University, Ottawa, Ontario K1S 5B6, Canada

Eleonore Kofman, School of Geography and Planning, Middlesex Polytechnic, Queensway, Enfield, Middlesex EN3 4SF, U.K.

Alexander B. Murphy, Department of Geography, University of Oregon, Eugene, Oregon 97403-1218, U.S.A.

Juval Portugali, Department of Geography, Tel Aviv University, Ramat Aviv 69 978, Tel Aviv, P.O.B. 39040, Israel.

Graham Smith, Department of Geography, University of Cambridge, Cambridge CB2 3EN, U.K.

H. van der Wusten, Subfaculty of Social Geography, University of Amsterdam, Jodenbreestraat 23, 1011 NH Amsterdam, The Netherlands.

Colin H. Williams, Department of Geography and Recreation Studies, North Staffordshire Polytechnic, Stoke on Trent, Staffordshire ST4 2DF, U.K.

222

PAPERS PREPARED FOR THE SAN SEBASTIAN CONFERENCE
(* not presented)

Session 1 Some General Issues (Chair; R.J. Johnston)

* M. Mikesell, Department of Geography, University of Chicago, IL 60637, U.S.A.
  When Subordinate Groups become Insubordinate: Reflections on the Causes of Ethnic Discord in Modern Nation States
  H. van der Wusten, Subfaculty of Social Geography, University of Amsterdam, 1011 NH Amsterdam, Joden-breestraat 23, The Netherlands.
  Rise, Persistence and Decline of Nationalist Movements.
  S. Waterman, Department of Geography, University of Haifa, Mount Carmel, Haifa 31999, Israel.
  Partition, Disintegration, and the Restricting of States.
* J. Ossenbrugge, Geography Institute, Hamburg University, 2 Hamburg 13, Federal Republic of Germany.
  Defending Space by Developing Territorial Ideologies.

Session 2: Mediterranean Perspectives (Chair: E. Kofman)

A. Drysdale, Department of Geography, University of New Hampshire, Durham, New Hampshire 03824, U.S.A.
  National Integration Problems in the Arab World
N. Kliot, Department of Geography, Haifa University, Haifa 31999, Israel.
  Nationalism, State-Idea, Self-Determination and War in the Mediterranean Region
D. Newman and J. Portugali, Department of Geography, Tel Aviv University, Ramat Aviv 69 978, Tel Aviv, Israel.
  Geography and Nationalism: Israel and the West Bank.
M. Roman, Department of Geography, Tel Aviv University, Ramat Aviv 69 978, Tel Aviv, Israel, Israel.
  Territorial versus Demographic Control as Competing Goals: The Case of Jewish-Arab National Struggle.
Y. Gradus, Department of Geography,Ben Gurion University of the Negev, Beer Sheva 84 120, P.O.B. 653, Israel.
  The Emergence of Regionalism in a Centralized System: The Case of Israel.

Session 3: Further General Issues (Chair: B. Becker)

D.B. Knight, Department of Geography, Carleton University, Ottawa, K1S 5B6, Canada.
Self-Determination for Indigenous Peoples: The Context for Change
* P. Szeliga, Institute of Geography and Spatial Science, Polish Academy of Science, Krakowskie Przedmiescie 30, 00-927 Warsaw 64, Poland.
Economic Dependence as a Factor of International and Inter-Regional Differentiation
S. Brunn and G.L. Ingalls, Department of Geography, University of Kentucky, Lexington, Kentucky 40506.
Recent UN Votes on Political Independence and Self-Determination
C. Williams, Department of Geography and Recreation Studies, North Staffordshire Polytechnic, Stoke on Trent, Staffordshire ST4 2DF, U.K.
Minority Nationalism: Idealist Alternatives

Session 4: European Perspectives (Chair: D. Knight)

* A-L Sanguin, Department of Geography, Université du Québec à Chicoutimi, 930 rue Jacques-Cartier est, Chicoutimi, Québec, G7H 2B1
Minorité ethnique, Discrimination et Territorialité: le cas des Romanches en Suisse
A.B. Murphy, Department of Geography, University of Chicago, Chicago, IL 60637, U.S.A.
The Partitioning of Belgium along Linguistic Lines
E. Kofman, School of Geography and Planning, Middlesex Polytechnic, Queensway, Enfield, EN3 4SF, U.K.
The Retreat from Neonationalism: France in the 1980s
C. Hedges, Open University, Walton Hall, Milton Keynes, MK7 6AA, U.K.
Problems of Combining Labour and Nationalist Politics: the Case of European Regionalism since 1960
G. Smith, Department of Geography, University of Cambridge, Downing Place, Cambridge CB2 3EN, U.K.
The Soviet Multinational State and Centre-Periphery Relations in the Period of Advanced Socialism

Session 5: Third World Perspectives (Chair: G. Smith)

* I.H. Zaidi, Department of Geography, University of Karachi, Karachi-32, Pakistan.

On the Concept of State-Territorial Integrity: Emerging Pattern of Thought in Southwest Asia

J. Cobbah, Urban Morgan Institute for Human Rights, College of Law, University of Cincinnati, Cincinnati, Ohio 45221, U.S.A.
Towards a Geography of Peace in Africa: A Reexamination of the State and Self-Determination in Post-Colonial Africa.

A. Sanchez, Department of Geography, University of Concepcion, Concepcion, Chile.
Geographical Space and the Integration of the Chilean Territory

* C. de Castro Aguire, Universidad Central de Venezuela, Caracas, Venezuela.
La Autodeterminacion en la Perspectiva de la Geografia Comportomental.

* M.I.Siddiqui, Department of Geography, University of Juba, Juba, Sudan.
Nationalism and the Muslim World.

Session 6: Individual Case Studies (Chair: S. Brunn)

A. Burghardt, Department of Geography, McMaster University, Hamilton, Ontario, L8S 4K1, Canada
Marxism and Self-Determination: the Case of Burgenland

J. Anderson, Open University, Walton Hall, Milton Keynes, MK7 6AA, U.K.
Nationalist Ideology and Territory

D. Mercer, Department of Geography, Monash University, Clayton, Victoria 3168, Australia.
Australia Aboriginal Land Rights: The Demise of a Vision

B. Becker, Universidade Federal do Rio de Janeiro, Brazil, South America.
Regionalism in Brazil

P. George, Department of Geography, Carleton University, Ottawa, Ontario, K1S 5B6, Canada.
The Gulf Cooperation Council

INDEX

absolutism   22-3
Africa   7, 70-86
Algeria   89, 193, 199
Amazonia   40, 43, 44, 51-3, 54-5
Arab world   7, 87-101, 193
Australia   8
Austria   58-69
authoritarian   9

Basque country   2, 6, 25, 34, 102, 112, 177, 206
Belgium   135-50
boundaries   5, 6, 21-2, 36, 80
Brazil   6, 40-56
Britain   27, 32, 206
bureaucracy   8-9, 10, 130
Burgenland   57-69

Canada   8, 124, 129, 130
capitalism   9, 10, 12, 19, 27, 30, 31, 160
centralization   172-4
class   12, 26-30, 31, 35, 57, 95, 107-16, 130
class alliances   31-2
colonialism   10, 13, 22, 26, 27, 58, 70-3, 78
community   33-6, 212
consociational democracy   75-7
constitution   79
culture   33-6, 73, 76, 93, 205
Czechoslovakia   199

decentralization   80-2
development (economic)   41-3, 47-9, 80, 81, 130
dissenting minorities   118-9

economics   30-3
Egypt   88, 93, 193
England   22
ethnic groups   6, 71-2, 73-5, 76, 93-4, 120, 138-9, 166-88
ethnoregional societies   166-88

feudalism   19
France   6, 22, 27, 102, 193, 204, 205
frontier   43-9, 50, 58

gemeinschaft/gesellschaft   35-8
generative social order   154-7
geoethnicity   73-5, 78, 82-3
Germany   32, 61, 88
Ghana   74

global strategies   47-9
Guatemala   129

hegemony   7, 50
historiography   203-21
history   5, 7, 18, 111, 205, 209-12
human rights   118-9, 128
Hungary   27, 58-9, 61, 64-5, 199

iconography   90
ideology   11-12, 20, 23, 26, 41, 43-7, 97, 103, 104, 161,
        218
imperialism   32, 58, 109
implicate social order   153-4
indigenous people   117-34
industrialization   41-3
intelligentsia   176-81, 200, 205
internal colonialism   13, 34
international instruments   119-22
Iraq   89, 91, 94, 193
Ireland   24, 27, 29, 102-16, 189, 193, 194, 197
Israel   7, 24, 151-65
Italy   33, 88, 193

Jordan   90, 92, 93

kinship   73, 74, 97

land   125-7
language   6, 33, 93, 94, 120, 135-50, 204, 208-9, 216
Latvia   173, 175, 178-81, 182-3, 184
law   8, 72
Lebanon   87, 90, 93
liberal prescription   70-3, 77
liberation movements   58
Libya   89

Marxism   12, 57-69, 152, 159-60
Mauritania   199
Moloch   163-4
moral perspective   208-15
Morocco   88

nation   3, 7, 8-9, 10, 33, 105, 212, 216
nation-ness   181-4
nation-state   7, 10, 21, 22, 41, 53-4, 152, 154, 156
national identity   87-93, 98
national integration   87-101
national unity   93-9

nationalism
  aims of   190-1
  core doctrine   155
  defined   2, 190-3
  operations   191-2
  repertoires of action   192-3
  success of   11, 12, 189-202
  (and) territory   20-21
  theory of   12-13, 18,   151-65, 204, 208
  varieties of   2-3, 20
nationalist politics   102-16
nationalities question   62-4
Netherlands   198, 201
New Zealand   8, 126, 130
Northern Ireland   102-10

occurrences of nationalism   193-7
Oman   88

Palestine   7, 90, 92, 93, 151-65, 207
pan-Arabism   89, 91
'people'   121-5, 216, 217
place   70-86
Plaid Cymru   207, 208, 214-5
Poland   27, 199
power   5, 8-9, 35
privileges of place   174-5, 184

region   50, 111
regionalism   135-50
religion   6, 33, 59-60, 88, 93, 94-9, 104, 112, 113, 120,
          206-15

Saudi Arabia   93
Scotland   102, 177
SDP (Austria)   59, 62-4
secession   7
self   73, 93, 130
self-determination   2, 8, 57-69, 70-86, 117-34, 197
sense of place   215-8
social order   152-5, 205
social theory   151-65
socialism   9, 23, 24, 26, 59, 60, 73, 98, 103, 107, 108,
          110, 166-88
South Africa   29, 79, 193
sovereignty   127-30, 197
Soviet Union   9, 60, 166-88
space-time diffusion   155-7
Spain   6, 25, 102, 112, 193, 204
state   2, 3, 8-9, 10, 21, 25, 36, 43-7, 72, 74, 76, 170-2
sub-state nationalism   111-4, 120

success of nationalism   197-201
Sudan   80
Switzerland   138
Syria   7, 87-101, 193

territoriality   5, 21, 24, 25, 31, 81, 136-40, 169, 170-2
territory   3-8, 10, 18-20, 23-6, 49-50, 72, 73-5, 120, 216
terrorism   200
theory   12-13, 18, 34, 204, 208
Tunisia   88

United Nations   119-22, 123, 128
USA   125, 130
urban-rural   104-7
urbanization   44-5
Uzbekistan   175, 176, 178-81, 183-4

Vietnam   199

Wales   102, 206-15, 215-8
World Council of Indigenous Peoples   127-8

Yemen   88, 93

Zaire   199
Zionism   157-9, 161

Printed and bound by CPI Group (UK) Ltd, Croydon, CR0 4YY

22/10/2024

01777621-0010